How to Build Metal/Treasure Locators

No. 909
$7.95

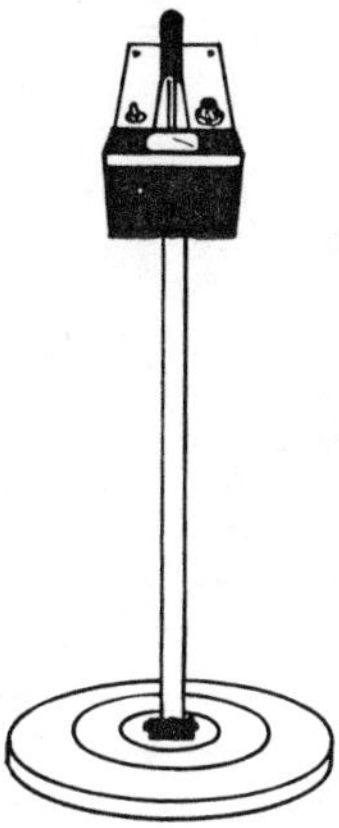

How to Build Metal/Treasure Locators

By Robert Traister
& John Traister

TAB BOOKS
Blue Ridge Summit, Pa. 17214

FIRST EDITION

FIRST PRINTING—MAY 1977
SECOND PRINTING—OCTOBER 1978

Printed in the United States of America

Library of Congress Cataloging in Publication Data

Traister, John E
How to build metal/treasure locators.

Includes index.
1. Metal detectors. I. Traister, Robert J., joint author. II. Title.
TK7882.M4T7 621.381 77-7510
ISBN 0-8306-7909-X
ISBN 0-8306-6909-4 pbk.

Preface

The purpose of this book is to supply the practical information needed to enable the reader to build his own metal detectors—regardless of his past electronics experience. The collection of wiring diagrams has been tested and proved to be reliable on "finds"—depending upon the circuit—from coins, to detecting electrical wiring in the home, and other metallic objects. All of the projects are designed for the hobbyist, requiring only the minimum tools, and with components that are readily available from any of the electronic supply houses.

The text is arranged in a logical sequence. The first chapter furnishes the reader with the fundamentals and basic techniques necessary to understand the construction and operation of all the detectors covered in this book. Brief discussions of the tools required are also covered to ensure a complete understanding of the construction techniques to follow. Each detector covered has a complete wiring diagram, a list of necessary tools needed for construction, complete material lists, and step-by-step instructions showing how to assemble the components into complete units. Finally, the text will tell exactly what each detector *will* and *will not* do, as well as give the reader helpful hints on the use of each detector for finding various kinds of "treasures". The tuning, adjusting, and servicing of each detector is also thoroughly covered.

All of the projects in this book—besides being fun to build—are valuable for their practical use. From the time you complete your first model, a successful treasure hunting hobby can be yours; by successful, we mean exciting and rewarding! No matter where you live—by sea or mountain, near desert sand or green forest—at least one of the detectors in this book will suit your needs and pocketbook. Then, with a little research, old coins, historical relics, antiques, old weapons, jewelry, and other valuables are yours for the finding.

The correct use of the detector is of the utmost importance to ensure constant "finds", but with the aid of this book, the reader will be able to understand the principles of metal detector operation, how to construct them, and how to operate them once they have been built.

John E. Traister
Robert J. Traister

Contents

1 The Essentials — 11
Workbench—Reading Schematic Diagrams—General Instructions

2 An Economical Metal Detector for the Beginner — 27
Parts—Assembly Instructions—Testing—Operation

3 Economy Model No. 2 — 43
Parts—Assembly Instructions—Testing—Operation

4 Self - Contained Detector — 57
Parts—Assembly Instructions—Testing

5 Miscellaneous Detector Circuits — 67
Substitute Components—Another Detector Using a Portable Radio—Metal Detector Meter Circuit—Another Self-Contained Detector—A Sensitive Detector—Another Detector Using A Meter

6 Building Metal Detectors From Kits — 81
Heathkit

7 **Repairs and Maintenance of Equipment** 89

Problems—(1) No Speaker Volume or Sound—(2) Poor Sensitivity or Detection Capability—(3) Excessive Frequency Shifting or Instability—(4) Erratic Operation—(5) Erratic-Sounding Speaker—Using Test Equipment for Troubleshooting

8 **How and Where to Use Your Detector** 97

Tuning Procedures—Field Operation—Digging Your Find—Choosing an Area

Appendices 103

A List of Equipment Dealers 105

B Selected Publications 121

C An Advanced Project 129

Index 137

Chapter 1
The Essentials

Metal detectors had their beginning almost fifty years ago when Dr. Gerhard Fisher, at that time a recent engineering graduate of the University of Dresden, arrived in Palo Alto, California, and went to work for the old Federal Telegraph Company. He was assigned to development work on radio compasses and direction finders, working with Dr. Frederick Kolster. When Federal moved to New Jersey, Dr. Fisher continued his direction-finder work with Western Air Express in Los Angeles. During this period, he observed that the bearing errors in radio direction finders were caused by large metallic structures, such as buildings and ore bodies which lay in the path of the radio waves.

Dr. Fisher perceived the possible usefulness of this kind of error for the location of ore bodies, and this led him into geophysical work, where he applied the method to mineral, oil and related geological surveys. Early in the depression years, he moved back to Palo Alto and in 1931 founded the Fisher Research Laboratory in his garage. Here, he continued to work both on direction finders and geophysical prospecting instruments using the radio principle.

His capability attracted the attention of the U.S. Navy and Dr. Fisher was asked to develop a direction finder for the dirigible U.S.S. Macon, together with a homing beacon for use by the small fighter aircraft carried by the huge airship.

Regrettably, the huge airship was lost in a storm off the coast of California in 1935.

By this time, the development of metal-mineral detectors had progressed to the point where Dr. Fisher applied for a patent on the instruments, which he called the "Metallascope", later shortened to "M-SCOPE".

The patent was issued in 1937, but meanwhile the detectors being manufactured were literally snatched up by public utilities like water and power companies, geophysical experts, and serious prospectors.

WORKBENCH

Before you begin constructing any of the metal detectors, there are several techniques you should fully understand; they are not difficult, so you should be able to learn them in a very short period of time. You should also obtain the necessary tools and have a suitable place to work.

An elaborate workbench is not necessary—a folding card table set up near an electrical outlet will be fine, but do not use a table with a metal top unless provisions are made to cover the metal portion with some sort of nonconductor, such as a piece of cardboard, plywood, or similar material. A metal-top table could cause short circuits if the wires are allowed to come into contact with the surface, and that could damage the components used in the circuits.

TOOLS

The tools needed to build your metal detector may be limited to a pair of needle-nose pliers, a pair of diagonal wire cutters, a pair of general-purpose pliers, a small soldering iron, a small amount of good quality resin-core solder, two blade and two Phillips-head screwdrivers (one small and one medium size). A pocket knife and a 1/4-inch drill motor (electric drill) with an assortment of bits will complete the basic tool list.

Other tools that may come in handy would include a round metal-cutting die 1/2-inch in diameter, a bench vise metal-cutting file, and a standard AM broadcast radio for monitoring detector frequencies. You will also need a selection of small bolts, nuts, and screws for the metal work involved.

Chances are, you already have these tools on hand, but if you are short one or two of the items mentioned, you can buy them from hardware stores, radio-supply houses, or mail-order firms. You probably don't need to be told that good tools are a worthwhile investment, so don't hesitate to select tools of high quality when purchasing those you need.

Needle-Nose Pliers

It was mentioned previously that you should have three pair of pliers for constructing the metal detector projects in this book. Of the three, the long nose or needle-nose pliers will probably be the ones most often used. See Fig. 1-1. They will be used to hold wire in place for soldering, to remove wires, or to grasp small items in hard-to-reach places. You should never use needle-nose pliers to tighten nuts, as you may spring the jaws to a point where they will not meet or close properly. However, the needle-nose pliers may be used to hold a nut in position so that it can be started on a bolt.

Diagonal Cutting Pliers

The pair of diagonal cutting pliers shown in Fig. 1-2 are the type used for most cutting operations and they should never be used for any other purpose. Used improperly, you will not only ruin the pliers, but mar the material on which they are being used as well. Notice that the cutting jaws are at an angle for cutting wires close to terminals.

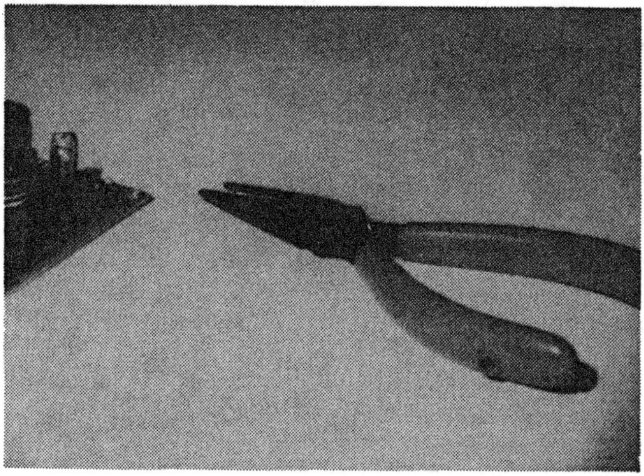

Fig. 1-1. Needle nose pliers.

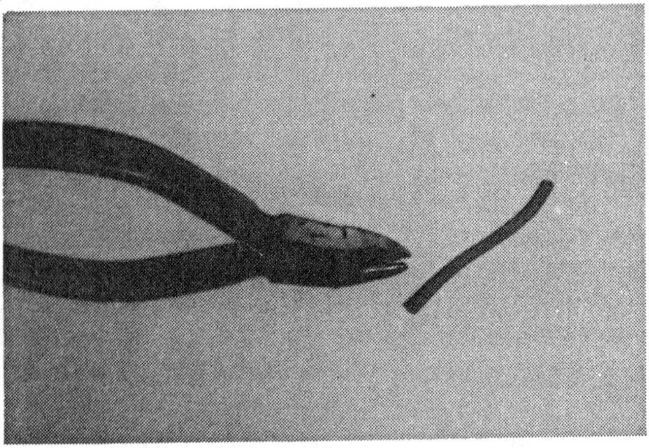

Fig. 1-2. Diagonal cutting pliers.

All-Purpose Pliers

The pliers shown in Fig. 1-3 are also handy for the projects that you will encounter. Because of the slip joint (another name by which these pliers are sometimes called), the jaws can be opened wider at the hinge pin so that larger items can be gripped. Although these pliers come in many different sizes, the thin 6-inch type will be the best for work on the projects in this book.

Screwdrivers

Everyone is certainly familiar with the standard screwdriver used to loosen or tighten screws. See Fig. 1-4. It is also probably one of the most abused tools in your kit because many people use a screwdriver as a pry or pinch bar to open a can of paint, and sometimes, even as a chisel. Use the

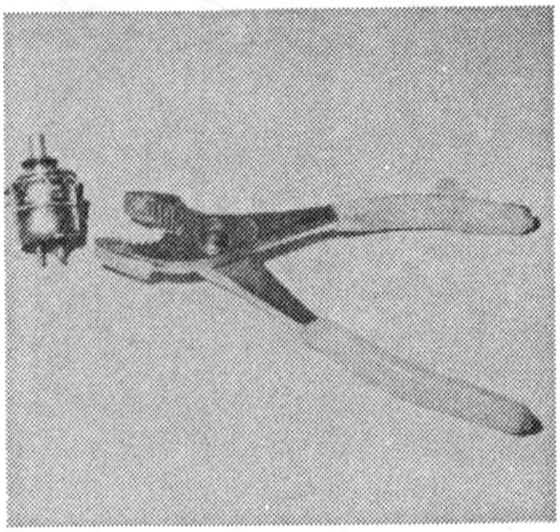

Fig. 1-3. All purpose pliers.

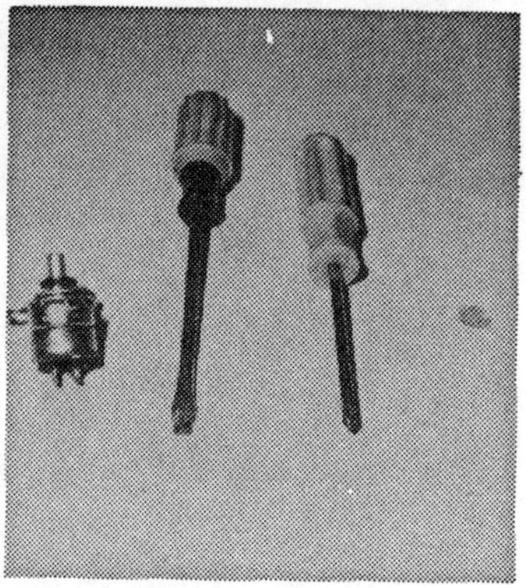

Fig. 1-4. Standard and Phillips screwdrivers.

screwdrivers in your tool kit only to loosen and tighten screws; used this way they will last you a lifetime.

You should have one small-blade screwdriver for loosening or tightening setscrews in dial or control knobs and one with a medium blade tip for general purpose work. You will also need a small and medium size Phillips screwdriver for use with screws with a special head known as the "Phillips Head." All of your screwdrivers should have insulated plastic handles to prevent possible electric shock should the blade come into contact with a "live" wire of sufficient voltage to shock you.

Soldering Iron

An inexpensive 40-watt soldering iron as shown in Fig. 1-5 will work very well on any of the projects in this book. However, if you anticipate doing other work in electronics, a soldering gun—like the one shown in Fig. 1-6—would be a better choice. This soldering tool combines all the advantages of a light-weight pencil type iron, a fast heating soldering gun, and a tip temperature control. The one pictured weighs only 7

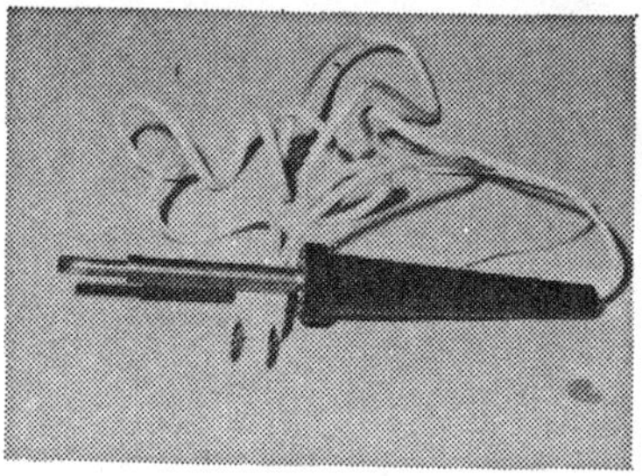

Fig. 1-5. An inexpensive 40-watt soldering iron that will work very well on any of the projects in this book.

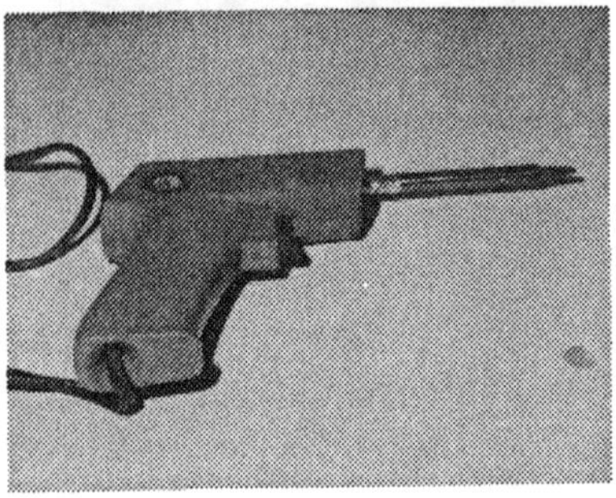

Fig. 1-6. A light weight pencil type soldering gun for fast heating.

ounces and is equipped with a 600°F, 1/8-inch conical point, soldering tip. The trigger turns the heat on and off and is ideal for electronic work, models, and other hobbies.

While we're on the subject of soldering tools, it would be wise to discuss the procedures for using these tools. The majority of home projects that fail do so because of improper soldering techniques.

In general, the materials to be joined should be clean and free from grease. If the terminals of the wires are not bright, scrape them with a pocket knife as shown in Fig. 1-7 or else with a piece of fine sandpaper (Fig. 1-8) until the ends are clean and bright.

Always apply the soldering tip to the joint to be soldered—not the solder itself; when the joint is hot enough the solder—also applied to the joint on the opposite side from the soldering tip—will flow onto the joint, leaving a shiny film when cool. A dull or jagged looking solder joint is an indication of an improperly soldered joint and is certain to present problems in the future.

Fig. 1-7. If the terminals of component wires are not bright, scrape them with a pocket knife, or...

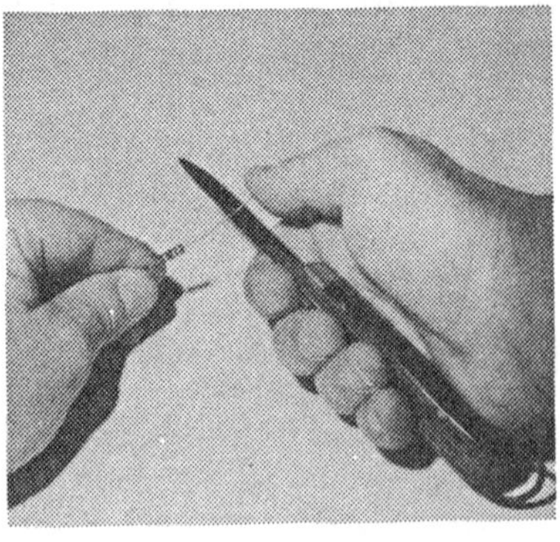

If these basic rules are followed, you should never have soldering trouble. If they are ignored, you may spend hours looking for defective components when the trouble is probably a poorly soldered joint. Soldering techniques will be discussed in greater depth later on in this chapter, but let's continue on to the remaining tools at this time.

Electric Drill

A small 1/4-inch hand drill is invaluable for drilling small holes in circuit boards and their housings. You can purchase a relatively good tool with all kinds of attachments for under $25. You should also have an assortment of drill bits, 1/4-inch and smaller. One kit on the market offers 13 high-speed drill bits in sizes from 1/16-inch to 1/4-inch for approximately $6. The kit also includes a center punch which assures exact positioning of the drill bits on the material to be drilled.

Metal Cutting Dies

Metal cutting dies or chassis punches are available in sizes from 1/2-inch to 3 inches or more; these will cut holes in metals up to 16 gauge thickness with the turn of an ordinary wrench. You will find the round punches the most useful although there are types for punching square and keyed sockets also.

Knife

A sturdy pocket knife is useful when preparing (scraping) wires for connection to other components. One of the blades should not be too sharp to avoid nicking the wires as they are being scraped.

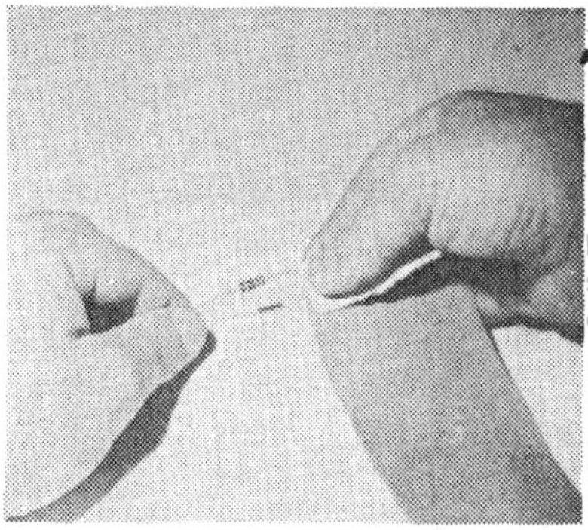

Fig. 1-8. ...use a piece of fine sandpaper until the ends are clean and bright.

Bench Vise

A small bench vise with jaws no larger than 4 inches will come in handy for holding parts while work is being done on them. Purchase either a portable vacuum-base vise, which is held in place on any non-porous surface by vacuum power, or a portable clamp type that is secured by a screw clamp to the edge of a bench or table.

There are other tools that will prove beneficial to your work, but the ones mentioned should get you through any of the projects listed in this book. If additional tools are needed for any of the projects, they will be mentioned in the chapter covering that particular project.

READING SCHEMATIC DIAGRAMS

To build any of the projects in this book, you must be able to know the meaning of the symbols used to indicate the various components and their related connections; that is, symbols are used to indicate the various parts and components, while lines are used to show the connections between them. This is known as a schematic drawing or diagram. You must remember that when working with schematic diagrams, the electrical connections are shown and not the actual physical connections.

Symbols

The symbol list in Fig. 1-9 should be studied carefully until you are familiar with the meaning of each. There aren't many, and this won't take up much of your time. Additional symbols used in the wiring diagrams will be explained in the chapter covering the diagram.

Now let's take each of the symbols in Fig. 1-9 individually and see what each consists of, and also what relation each has to metal detector circuits—especially the ones found in this book.

Resistor. An electronic component that offers resistance to the flow of electric current. The various electrodes in transistors used in metal detector circuits do not all require the same operating voltages although all obtain their voltages from a single power supply. Therefore, resistors are used to drop the voltage to the correct value for operation of the transistors.

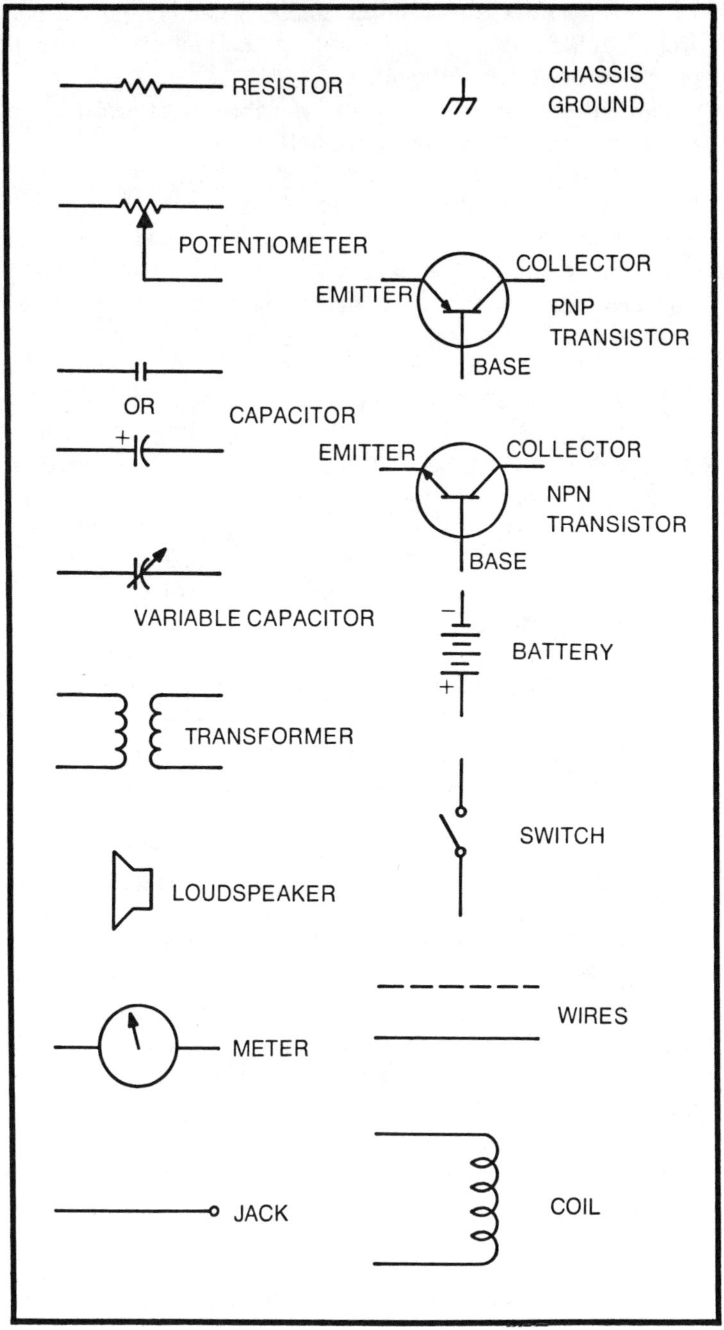

Fig. 1-9. Electronic symbols used in schematic diagrams.

Resistors are also used to isolate circuit component parts from each other, so that one will not interfere with the operation or action of the other.

Potentiometer. These are variable resistors used to control the volume or tone of the detector signal.

Capacitor. Capacitors are used in electronic circuits to block the flow of direct current, and to allow the flow of alternating current to a certain degree, depending on the capacitance and the frequency. Most consist of two conducting surfaces separated by insulating materials of various substances.

Transformer. A device consisting of two or more coils constructed in such a way that magnetic lines of force produced by the flow of alternating or pulsating direct current through one coil will pass through the other coil and induce in it a corresponding alternating voltage.

Loudspeaker. A device used to radiate acoustic power effectively at a distance through the air. When a loudspeaker is incorporated into a metal detector circuit, it usually is intended to reproduce and magnify the tone from the oscillator.

Meter. Those used in metal-locator circuits normally consist of the d'Arsonval type with a range of 0–50 DC milliamperes. A change in the meter reading indicates a "find," the same as a change in tone from the loudspeaker or earphones.

Jack. A plug-in spring terminal widely used in electronic circuits for temporary connections of batteries, earphones, etc.

Coil. A number of turns of insulated wire offering considerable opposition to alternating current at the frequency for which it is designed. A coil is used in the search head of metal detectors in a similar way that an antenna coil is used in a radio; that is, a receiver circuit of coiled wire through which the antenna current flows.

Ground. A "ground" used in metal detector circuits usually refers to the chassis of a receiver serving as a common path for signal circuits.

PNP Transistor. A transistor in which the base is made up of semiconductor material that has been doped with a donor material that gives an excess of electrons, while the emitter and collector are made of semiconductor materials that have been doped with an acceptor material that gives an excess of holes.

NPN Transistor. A transistor in which the material arrangements are in reverse of a PNP transistor.

Battery. A device consisting of one or more dry cells or storage cells connected together to serve as a DC power source in metal detector circuits.

Switch. A device used to open and close an electrical circuit—like from the battery power supply—or for changing the connections between parts or circuits. When used for the latter, it is commonly known as a transfer switch.

Wires. Wires are used to connect the various components and are shown in the schematic diagrams by means of lines. Where two lines cross and a dot is placed at the junction, this means the wires are to be connected or soldered at this junction; otherwise, they merely cross each other and are not electrically connected.

GENERAL INSTRUCTIONS

The basic parts of a typical metal locator are shown in Fig. 1-10. The search or pickup coil consists of several turns of insulated wire. This coil should be wound on a mechanically stable form such as a cardboard Tinker Toy box or similar object that is not subject to bending or warping. The wound coil is then secured with epoxy cement to some type of housing. A plastic Frisbee or several stiff paper plates stacked on each other and then varnished make excellent homemade forms. A hollow plastic wagon wheel from a child's discarded toy is another possibility.

An aluminum tube is secured to the search coil housing (other materials will work just as well) and the oscillator control box is normally secured to the end opposite the search coil. As little metal as possible should be used in connecting the coil housing to the tube for best results; some hobbyists prefer to use plastic screws for this connection.

The finest metal detectors on the market today are those which boast not only advanced circuitry but advanced mechanical stability as well—both are equally important. Therefore, you must make certain that all components and connections are firmly secured, because poor mechanical stability will result in drift of the oscillator frequency. Secure, rigid mounting of all parts can be obtained by using short lead

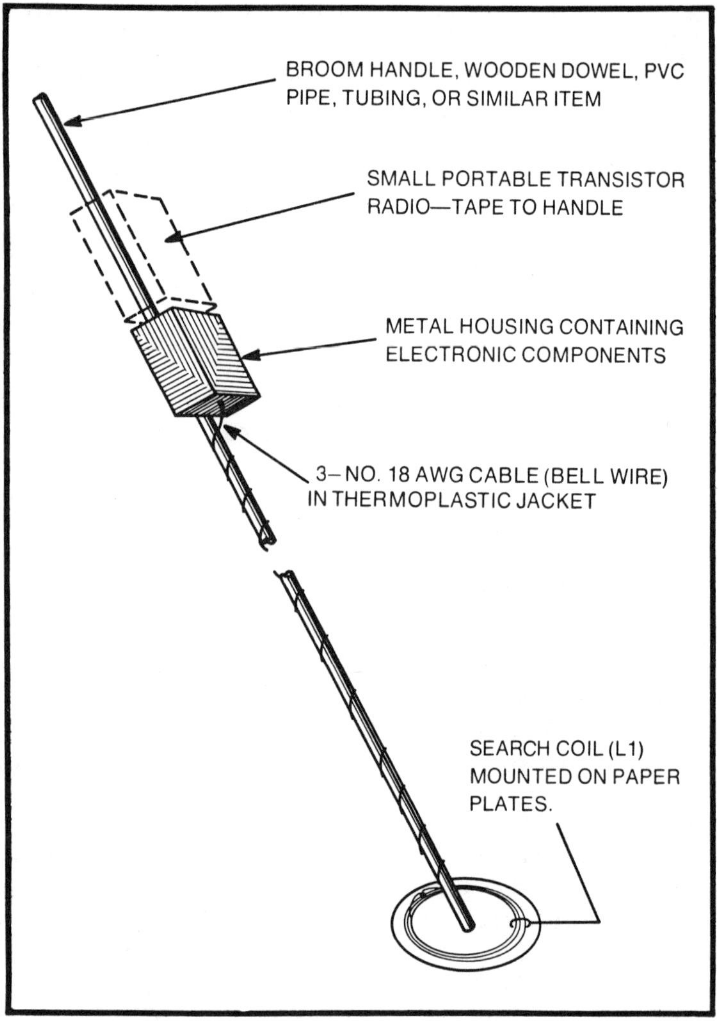

Fig. 1-10. The basic parts of a typical home made metal locator.

lengths on the components and by using epoxy cement to hold fast any component that tends to shake or move in any way when the unit is moved. After your oscillator unit is completed, give it a good firm shake, while observing the circuit components. If anything moves, dab several drops of epoxy on the guilty component and let set until dry and hard.

Many who have built metal detectors prefer to complete the mechanical work first; that is, the oscillator and coil housings, the tube handle, etc. The electronic wiring may seem to be the more difficult of the two, but the presence of a completed mechanical shell to place the electronic components in when they are completed could be an added impetus to keep you going—should you get frustrated halfway through—instead of possibly giving up entirely.

Soldering Techniques

Earlier in this chapter the importance of good connections was discussed as well as the basic steps involved in soldering these connections. Now these steps will be covered in more detail to ensure that your connections will be trouble-free—allowing you more pleasure for a longer period of time looking for treasure and not for troubles with your detector.

First of all, all leads of the components or connecting wires should be tinned before you attempt to solder them. While most of the components purchased from reliable electronic supply houses have been tinned by the manufacturers, the tinned surface sometimes becomes covered with wax or other impurities requiring cleaning and retinning.

Clean the lead with your pocket knife, and then hold the component with your long nose (needle-nose) pliers and touch it to the tip of your soldering iron as shown in Fig. 1-11. Then touch a piece of solder to the lead and move the lead back and forth through the solder to tin the entire lead. Enough heat should be applied to the lead so that the solder will flow smoothly over the lead. Then tin the opposite lead in the same way.

To solder a lead to a terminal strip or solder lug, place the lead through the opening in the terminal lug, bend the lead slightly so that the lead can be placed in contact with the metal part of the lug. Do not wrap the lead around the terminal strip

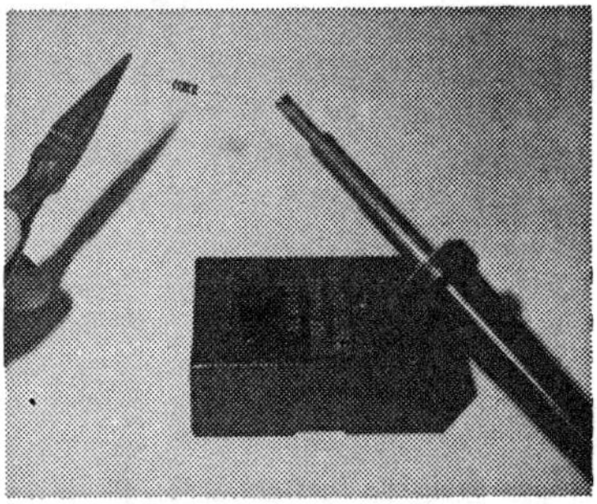

Fig. 1-11. Method of tinning leads of components prior to connection.

lug unless you find it absolutely necessary to hold it in place while soldering; this type of connection is too difficult to move if a mistake is made, or you wish to modify the circuit later. However, if you are certain of your connections, and you will not desire to change them later, wrapped connections will help to ensure strong mechanical and electrical connections.

With the lead in place, hold your soldering iron against the terminal and against the lead to heat both of them to the proper temperature. Then touch the end of the solder to the terminal and lead, so that the solder will flow freely over the lead and the terminal. Do not use too much solder, as it takes only a drop of solder to make a good joint. Once you have one drop of solder flowing around the joint, lift the solder away from the joint, but leave the soldering iron on the joints so that the solder on the terminal will flow around the leads and over the terminal. If you hold the soldering iron on the joint only long enough to melt the solder and have it start to flow, you will find that you will have a rough-looking joint, with the possibility of the leads pulling loose under the slightest pressure.

Color Codes

Let's take a look at one more item before you begin constructing the metal detector of your choice—the indentification of resistors. Although the value is sometimes stamped on the resistor itself, most one-half through 2-watt resistors are identified by colored bands on the resistor. You will need to be able to identify the value of these resistors for the projects described in this book.

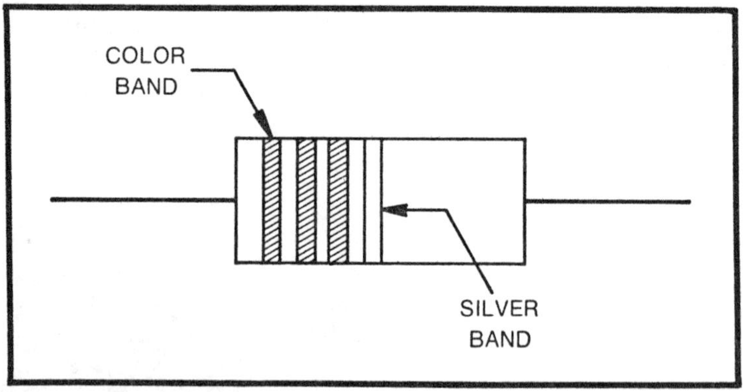

Fig. 1-12. To read the color code of a resistor, position the resistor so that the colored bands are towards the left end of the resistor as shown here.

To read the color code, turn the resistor so that the colored bands are toward the left end of the resistor as shown in Fig. 1-12. On the resistor pictured, the colored bands are nearer to one end than the other. However, on some types of resistors, the colored bands are centered; in the latter case, place the gold or silver band to the right, and you will have no trouble reading the code.

Each color represents a number as follows:

Black	0
Brown	1
Red	2
Orange	3
Yellow	4
Green	5
Blue	6
Violet	7
Gray	8
White	9

Therefore, the resistor in Fig. 1-20 with the colors (from left to right) yellow (4), violet (7), and brown (1) is read as follows:

The first yellow band represents the number 4 and the second band represents the number 7; the two combined colors gives us a value of 47. The third color (brown) gives the number of zeros in the value. Since brown represents 1, the third band indicates that the resistor has 1 zero in its value. So the value of the resistor is 470 ohms.

If the resistor had been color-coded yellow, violet, and red, the first two figures would again have the total value of 47 (yellow = 4, violet = 7). Red represents 2, so there would be two zeros, and the value would be 4700 ohms—often written 4.7K ohms as the "K" stands for 1000. If the colors were yellow, violet, orange, the value would be 47000 ohms, etc.

When you purchase the resistors for your detector project, look them over, and practice reading the color codes on them.

With this basic data out of the way, you should have no trouble following the step-by-step instructions given for the construction of metal detectors.

Chapter 2

An Economical Metal Detector for the Beginner

You can locate buried plumbing pipes in your backyard, electrical wiring inside of walls, find a lost ring or coin at the beach, or maybe even discover a cache of gold coins buried years ago by some miser who once owned your property, with the simple little metal detector as described here. It will detect the presence of metallic objects hidden by any nonconductive material such as dirt, wood, stone, plaster or plastic.

With less than $10 worth of electronic components and perhaps an hour or two of your time, you can be on your way to a successful treasure-hunting hobby. With a little practice, you will be able to detect metallic objects to depths of over several inches, and experience will even give you a fairly good idea of their size by listening to the change in audible pitch from the portable transistor radio which must be used in conjunction with this particular metal locator. This little detector uses the beat-frequency method to detect a "find", and the procedure works something like this:

The frequency of the search coil oscillator (see Fig. 2-1) changes when the conductance in the field of the search coil changes. Therefore, when metal or minerals come within a short distance of the coil, the frequency of the search coil is changed, which in turn changes the pitch of the audible sound. The larger the metallic object, the more noticeable the change in pitch; the smaller the object, the less noticeable the change

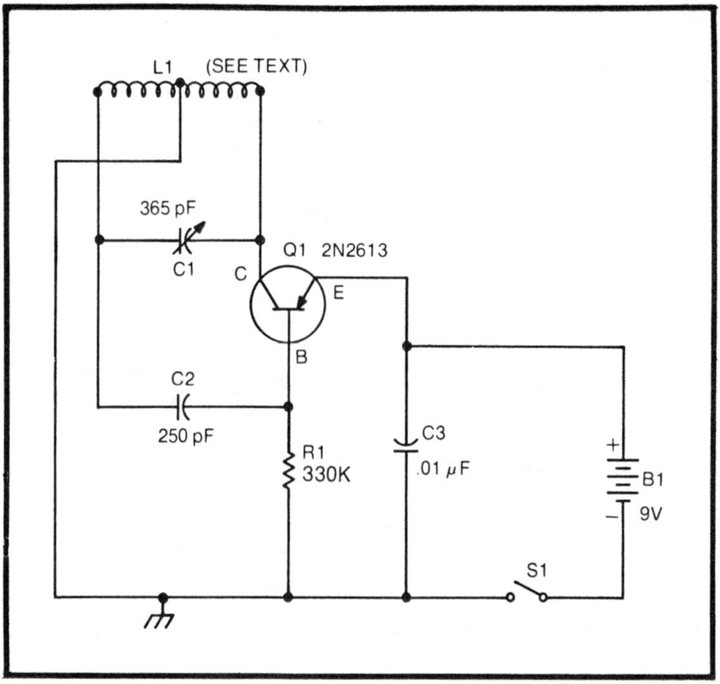

Fig. 2-1. Schematic diagram of the electronic components and their related connections.

in pitch becomes. In fact, for very small objects like coins and rings—at a depth of over an inch—you will have to practice considerably to be able to hear the change at all. On the other hand, larger items such as axe heads, steel traps, and the like, are easily detected your first time out.

This is a good project on which to begin, since it has less than a half-dozen electronic parts, and although you'll probably want to go on to a more sophisticated model later, this device has the capability of paying for itself over and over again. It is also a good model to practice treasure hunting with.

PARTS

To begin the project, you will need the following parts; the electronic items are easily obtainable at any electronics supply house or any TV repair shop:

1. One plastic Frisbee large enough for the 6-inch search coil to fit inside.

2. One 4-foot length of aluminum tubing with adjustable swivel on one end to secure the frisbee.
3. One bicycle handle grip to place on the other end of the aluminum tubing for gripping when operating the detector.
4. An aluminum housing approximately 3 inches wide × 5 inches long × 2 inches deep for mounting the switch, trimmer, battery, and circuit board.
5. Tape, miscellaneous screws and nuts, and epoxy cement.
6. One 9-volt transistor battery holder with mounting screws.

Refer to the schematic wiring diagram in Fig. 2-1 and the pictorial wiring arrangement in Fig. 2-2 in order to better visualize the following electronic components.

7. A circuit board approximately 2 1/2 inches × 2 inches.
8. L_1—20 turns of 22 AWG enameled magnet wire wrapped on 6-inch form and tapped at 10 turns. This will require approximately 23 feet of wire.
9. C_1—365 pF variable capacitor for tuning the transistor radio signal so that it "beats" against the received signal, which will produce a whistle sound in the receiver.
10. C_2—250 pF mica capacitor
11. C_3—.01 μF paper capacitor

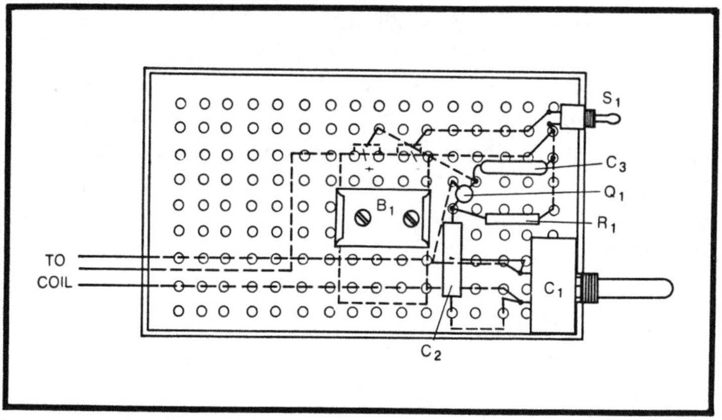

Fig. 2-2. The physical arrangement of the electronic components on the circuit board.

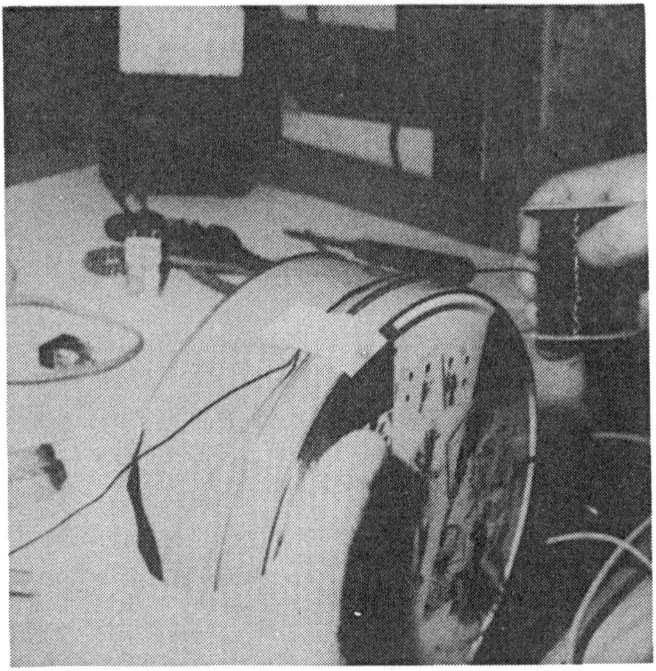

Fig. 2-3. Tape one end of the magnet wire to the winding form leaving approximately 8 to 10 inches of lead wire before beginning the number of turns required. Metal detector coils are wound in scramble fashion (which means don't be neat).

12. Q_1 —2N2613 PNP transistor (Radio Shack 276-2003 or equal)
13. R_1 —330K 1/2-watt carbon resistor
14. S_1 —single-pole, single-throw switch
15. B_1 —9-volt transistor radio battery

Also needed are a few feet of hook-up wire, solder, and a three-wire cable long enough to reach from the search coil at one end of the handle to the circuit board at the opposite end of the handle. Number 18 to 22 AWG bell wire will work, but shielded cable will give better results. This should complete the list of components, but you will need the tools described in Chapter 1.

ASSEMBLY INSTRUCTIONS

As you construct the project, make certain that the diagrams Figs. 2-1 and 2-2 are in sight at all times for easy reference, and double-check all of your connections before soldering them.

Winding The Coil

The coil is going to require some waiting time for the epoxy cement to dry, so this step would be the logical one to begin first. Then while you're waiting for the cement to dry, you can construct the circuit board—wiring the various components.

Begin by securing one end of the enameled magnet wire to the 6-inch form with masking tape as shown in Fig. 2-3. Note in this figure that a metal cake tin was used for the form. A cardboard, plastic, or wooden form would have been safer as the magnet wire scraping against the metal could cause some of the insulation to be removed from the wire. Therefore, if a metal form must be used, use extreme care winding the coil and also remove it very gently from the form.

On the tenth wrap (halfway), form a loop in the magnet wire for your center tap, then continue until you complete the 20 turns of wire. See Fig. 2-4. These turns should not be wrapped in a neat manner; rather, they should be *scramble-wound* which means don't be neat. When completed, the scramble-wound coil should look something like the one in Fig. 2-5; that is, 20 turns of wire on a six-inch form with about ten inches of lead between taps. Collapse the form before

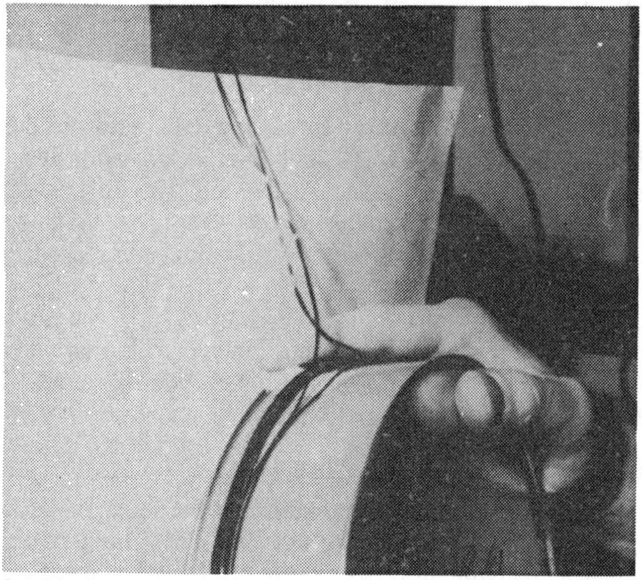

Fig. 2-4. Make a loop after the tenth turn as shown here for the center tap before continuing.

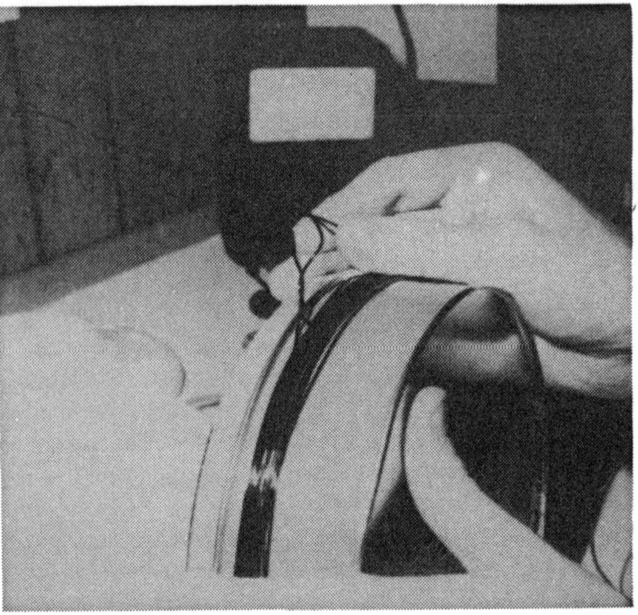

Fig. 2-5. The finished coil prior to removing it from the winding form.

removing the coil. In our case, the tin top was first removed from the cake tin, a slit was made with tin snips beginning from the rim then cutting in towards the coil. With the slit, the tin form could be collapsed so as not to injure the insulation on the wire.

Once removed from the form (Fig. 2-6), saturate the coil with coil dope or General Electric RTV coil adhesive. It is extremely important to secure the coil wire firmly, as one loose loop in the coil will cause the detector to be unstable. This is also true—to some extent—of all components used in the detector.

While you're waiting for the coil dope to dry, you can begin assembling the circuit board, but make certain that you have all of the parts and tools needed before you begin. Also have Figs. 2-1 and 2-2 in view at all times.

Mounting Components

The basic components required to construct this metal detector are shown in Fig. 2-7 and consist of (from left to right) a 9-volt transistor battery, a spool of 22 AWG magnet wire, a fiber circuit board, a variable capacitor, epoxy cement in

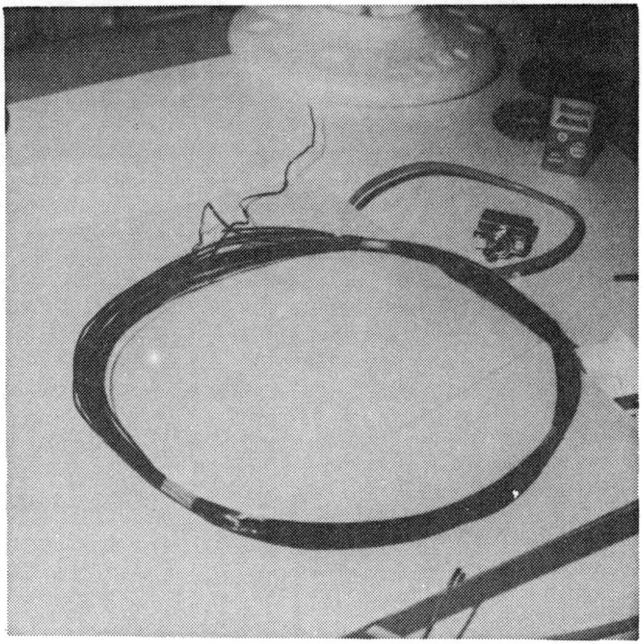

Fig. 2-6. The finished coil after it was removed from the winding coil.

applicator, and a spool of solder. The switch, capacitors, transistor, and resistor complete the components.

First arrange the capacitors, the transistor holder, and the resistor on the circuit board as shown in Fig. 2-1. Only twist the

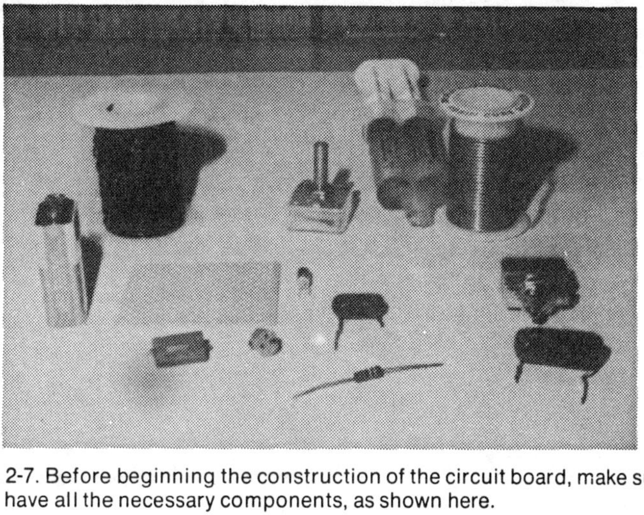

Fig. 2-7. Before beginning the construction of the circuit board, make sure you have all the necessary components, as shown here.

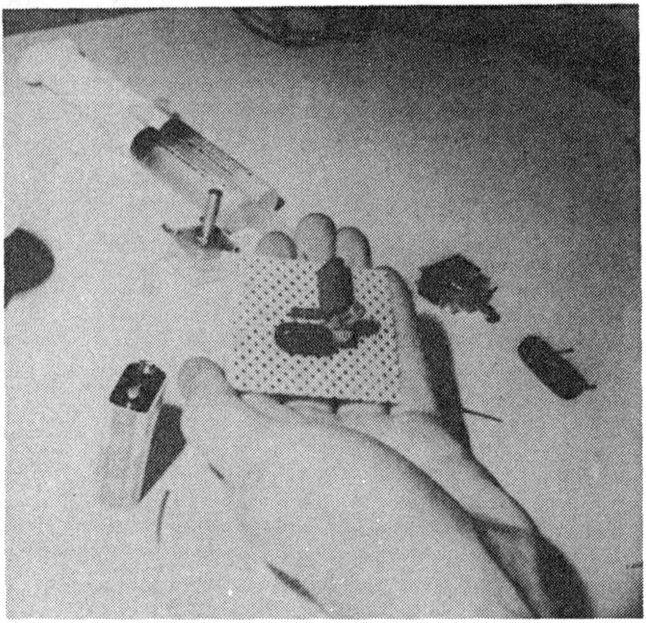

Fig. 2-8. Appearance of the circuit board once the basic components have been installed.

Fig. 2-9. Method of soldering the connections on the bottom of the circuit board.

connections together at this time—do not solder. Your circuit board should now look similar to the one in Fig. 2-8. Next, connect the required hook-up wires, making sure they are long enough to reach their various connections which will be required later on. When you're certain that all twisted connections are correct, solder the connections as shown in Fig. 2-9. The bottom of your circuit board should now look similar to the one in Fig. 2-10.

Select a location for your battery holder, drill any required holes through the circuit board so the two screws securing this holder can be inserted through the board, and secure the battery holder to the circuit board using the screws and nuts normally supplied with the holder.

You are now ready to mount the variable capacitor (trimmer). Apply epoxy cement to the bottom of the trimmer as shown in Fig. 2-11 and attach it to the circuit board at the location shown in Fig. 2-1. Lay some type of weight across the trimmer to hold it in place while the epoxy is drying; a pair of pliers—like the ones in Fig. 2-12—make an excellent weight. The drying time for most types of epoxy is approximately 6 to 10 minutes.

While the epoxy is drying on the trimmer, let's get back to the coil; the coil dope should be dry by now. If you plan to use a

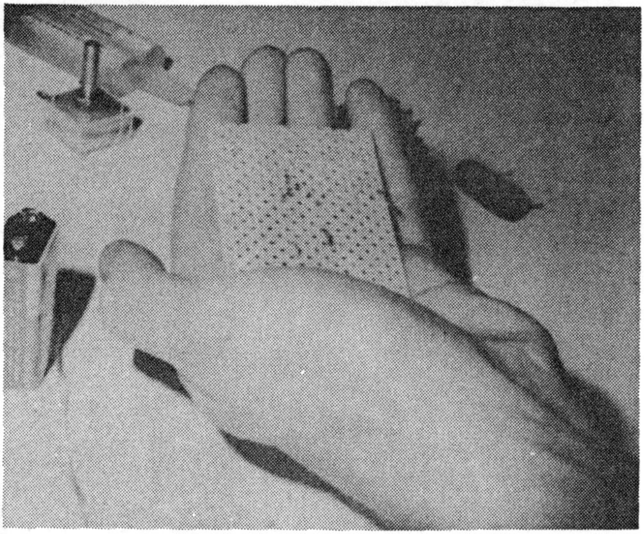

Fig. 2-10. The bottom of the circuit board should have a neat appearance after the soldering is completed.

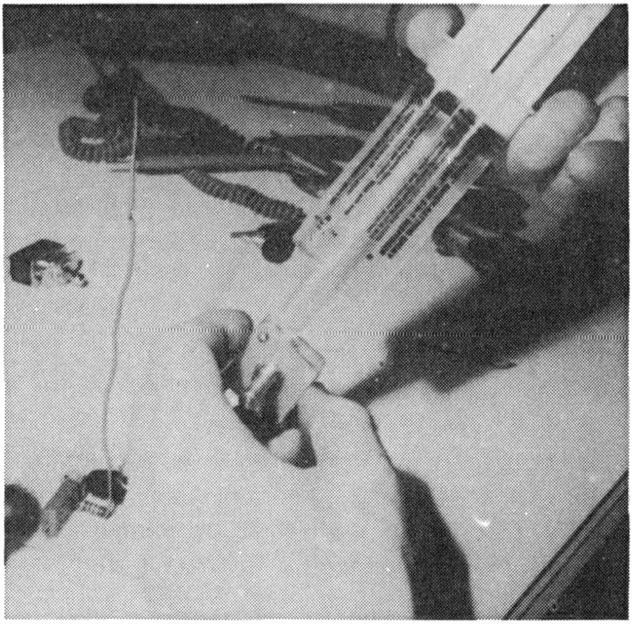

Fig. 2-11. Applying epoxy cement to the bottom of the trimmer using a special applicator that ejects both the agent and the hardener at the same time.

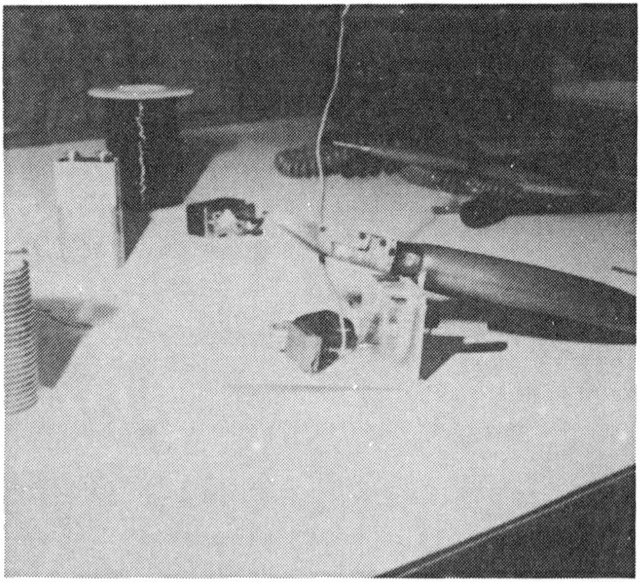

Fig. 2-12. A pair of cutting pliers are used to hold the trimmer in place while the epoxy is drying.

frisbee for the search head, the wound coil should be placed on the bottom side of the frisbee (the way you would throw it) and, therefore, three small holes must be made through the top of the frisbee to allow the three coil lead wires to pass through. Make the holes by heating the tips of each lead with a soldering gun, and with them melting the holes through the plastic. With the three leads now protruding through the top of the frisbee, secure the wound coil to the bottom of the frisbee with epoxy—setting some weight upon this coil while it dries in place.

You might as well get the remaining mechanical work out of the way at this time, so take the aluminum tubing, insert a wire-spring tubing bender over one end, and make a bend of approximately 60° far enough down from the end to slip the bicycle handle grip over the entire length of the bend. The aluminum housing can then be secured on the tubing directly beneath this bend, and when attached to the Frisbee search head, your detector should look something like the one in Fig. 2-13.

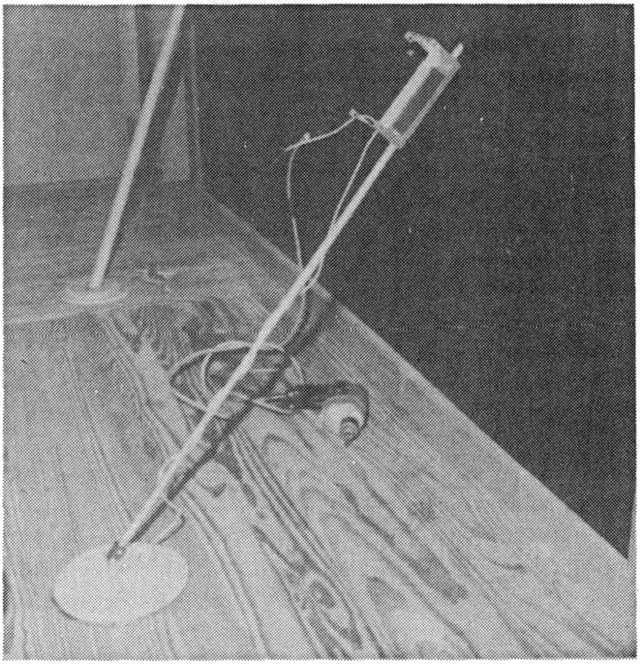

Fig. 2-13. Appearance of the mechanical parts of the metal detector; that is, the search coil, aluminum tubing handle, and aluminum housing.

Fig. 2-14. The inside of the aluminum housing after the circuit board, trimmer, battery, and switch have been installed.

Drill two openings large enough to accept the trimmer control shaft on one side of the aluminum housing, and another opening on the opposite side to accept the switch. Insert the switch, secure to the housing with the locknut, and then insert the circuit board into the housing. If the control shaft of your trimmer is metal, use a rubber grommet in the housing hole to prevent the shaft from coming into contact with the metal housing. Since the trimmer is firmly secured to the circuit board, this trimmer shaft—once the locknut is secured—will hold the circuit board in place. Just make certain that you drill the hole in the housing high enough so that the connections on the bottom of the circuit board will not come into contact with the metal housing once the trimmer locknut is secured. Figure 2-14 shows how the inside should look.

Connect the leads of the three-wire cable to the three leads coming from the coil on the search head—using a color code to identify each wire; that is, use the white (for example) wire of the cable for the center tap on the coil, another color wire for one of the end taps, and the blue (or whatever color it may be) for the tap on the other end of the coil. Then connect the other wire ends of this cable to their appropriate points on the circuit board.

If you don't happen to have a piece of aluminum tubing laying around your shop or home, an old broom handle or wooden dowel will work just as well for the detector handle. Conventional paper plates may also be used to house the wound coil, if you don't have a Frisbee of the right size. Take two or three paper plates, cement them together, and then apply two coats of clear varnish to all exposed sides of the paper-plate form. Cut the wooden broom handle or wooden dowel so that it is approximately 3 feet long, and then cut one of the ends at a 30° angle as shown in Fig. 2-15. You can then secure the wooden handle to the paper plates with epoxy cement as shown in Fig. 2-16. The wound coil is secured to the top (or bottom) of the paper plates with epoxy cement. Complete details of construction are shown in Fig. 2-17. The circuit board and all connections are made exactly as described previously and as shown in Figs. 2-1 and 2-2.

Check the inside of the housing (Fig. 2-14) once more to make certain that everything is in order before attaching the

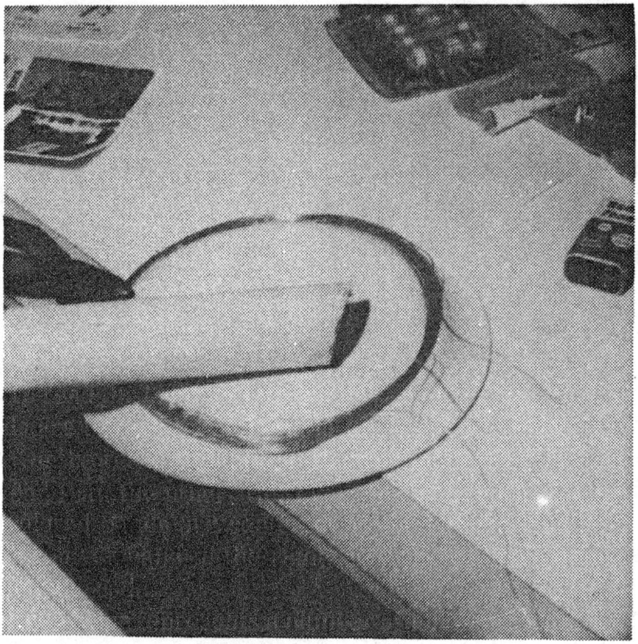

Fig. 2-15. Wooden dowel end cut at an angle of approximately 30 degrees.

Fig. 2-16. The wooden dowel attached to paper-plate search coil housing with wound coil attached.

housing cover. Then tape a small portable transistor radio just below the housing and on the aluminum handle. Use a good quality tape and make sure that your wraps are tight so the radio won't shake loose. You're now ready to test your creation.

TESTING

Take the detector outside, away from all metal objects, flip the switch on the metal housing, and then turn on the radio and tune to a local station. Adjust the trimmer on the detector circuit until you obtain a distinct whistle tone—usually occuring between two stations or on one side of a station. Pass the search head over a metallic object and the pitch of the whistle will change (usually higher). When the most pronounced change occurs, you should be directly over the metal object.

Like with any metal detector, you're going to have to play around with this one to obtain the best results; the more you practice, the more "finds" you will make.

OPERATION

The morning that this detector was first used, several metal toy soldiers were found buried approximately two inches under the ground around the perimeter of a residential front porch. Despite the damp grass which gave some "false" signals on the detector, we were able to locate dozens of buried nails and plumbing pipe up to 4 inches under the ground. We quickly learned, however, that this detector is not for the serious treasure hunter—but we hasten to add, it does work and will provide hours of fun for young and old alike finding the larger buried metallic items found around old homes.

Be sure to read Chapter 6 for hints on places to hunt as well as how to use your metal detector—regardless of the type.

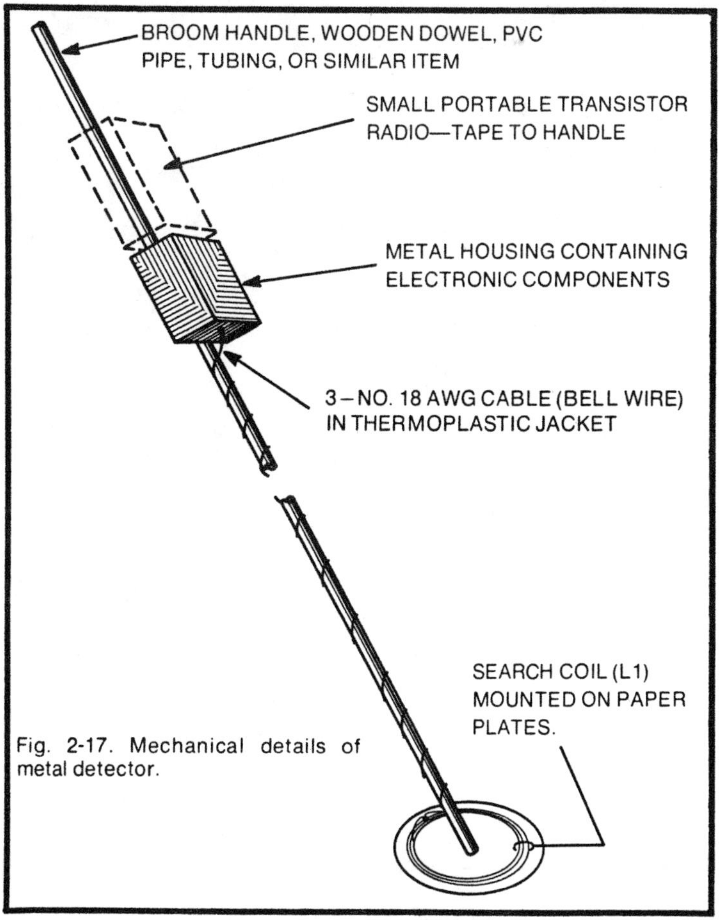

Fig. 2-17. Mechanical details of metal detector.

Chapter 3

Economy Model No. 2

Here's another beat-frequency type of metal detector which also uses a small portable transistor radio in conjunction with the circuit covered in this chapter. All parts are easily obtainable at your local electronic shop and the total cost of them should not exceed $12—providing, of course, you already have a small protable transistor radio. If you don't, this type of radio can be purchased for $4 to $6 at any of the chain department stores, drug stores or at most electronic supply houses. You can even obtain a kit from some of these sources and build your own.

If you're a beginner, the wiring and assembly of the electronic components should take approximately one hour to complete; experienced personnel should be able to complete the circuitry in only a few minutes. Once you have obtained all of the components, the necessary tools, and the hardware, the entire detector should not take longer than two hours to build. Then you're in for hours of fun using this device for locating metallic objects of all sizes: lost coins or rings; old metallic tools like axe heads or garden hoes; even locate electrical wiring and plumbing pipes in concealed walls or buried in the ground. For fun or profit, this inexpensive metal locator is in reach of everyone's budget and capabilities.

You will have no trouble assembling this metal locator if you follow directions carefully and perform one step at a time.

If this is one of your first electronic projects, be sure that you fully understand everything mentioned in Chapter 1 concerning construction procedures for metal detectors in general.

PARTS

The following parts and materials are needed to construct this detector:

1. One empty 18-oz. cardboard oatmeal box or similar circular object, for use as a form to wind the 4-inch diameter coil.
2. Approximately 20 feet of 22 AWG enameled magnet wire, to build a search coil consisting of 18 turns of wire, plus a few inches extra to reach to the circuit-board housing mounted close to the coil.
3. A broom handle or wooden dowel to be used as a mounting shaft and also a handle for operating the detector; that is, swinging the search coil back and forth when searching for objects.
4. Two or three paper plates approximately 6 inches in diameter on which to mount the 4-inch diameter search coil.
5. Clear varnish and a small paint brush to varnish the paper plates for added protection against dampness and also to make the plates more rigid.
6. Epoxy cement which will make a permanent bond in 4 to 6 minutes. (Used to secure some of electronic components as well as some of the mechanical items.)
7. Miscellaneous hardware such as nuts and bolts, wood screws, rubber grommets, electrical tape, etc.
8. Coil dope for saturating the wound coil to hold the turns in place.
9. Three 1/2-watt carbon resistors of the following values:

   47,000 ohm
   10,000 ohm
   680 ohm

10. One 365 pF variable capacitor for tuning the transistor radio signal so that it "beats" against the received signal—producing a whistle sound in the receiver.

11. Four fixed-value capacitors are necessary for this circuit: one 250 pF, 100-volt silver mica capacitor(two 0.05 μF) disc capacitors; and one 5 μF, 12-volt electrolytic capacitor.
12. One NPN transistor (RCA type SK3011 or Radio Shack 276-2001).
13. A few feet of assorted hook-up wire, solder, a circuit board, a 9-volt transistor battery, and a single-pole, single-throw switch should finish up the basic parts.

ASSEMBLY INSTRUCTIONS

When you have obtained all of the items on the parts list and have checked them over to make certain that everything you need is within easy reach, refer to the pictorial wiring diagram in Fig. 3-1; this shows you approximately how the physical layout of the components should look. Then refer to the schematic wiring diagram in Fig. 3-2 and compare the two.

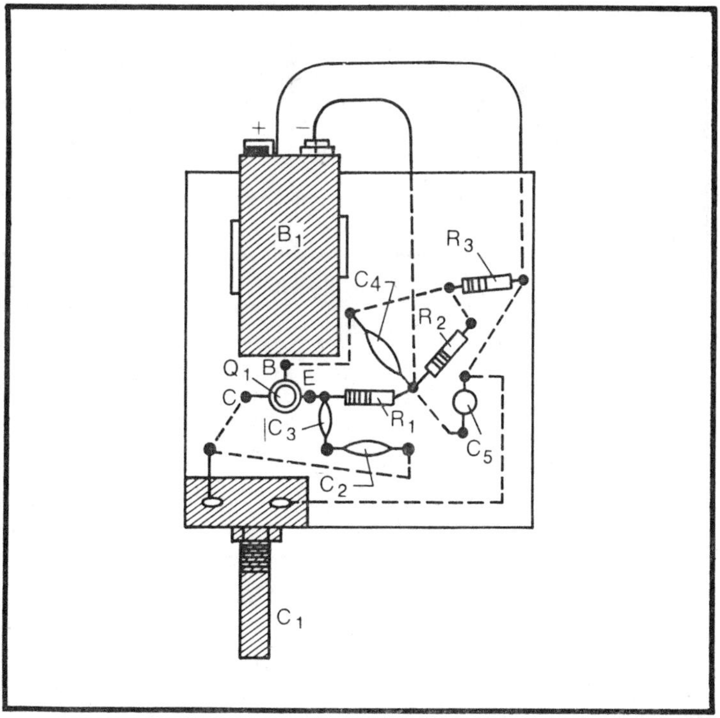

Fig. 3-1. Pictorial wiring diagram of the metal detector circuit described in this chapter

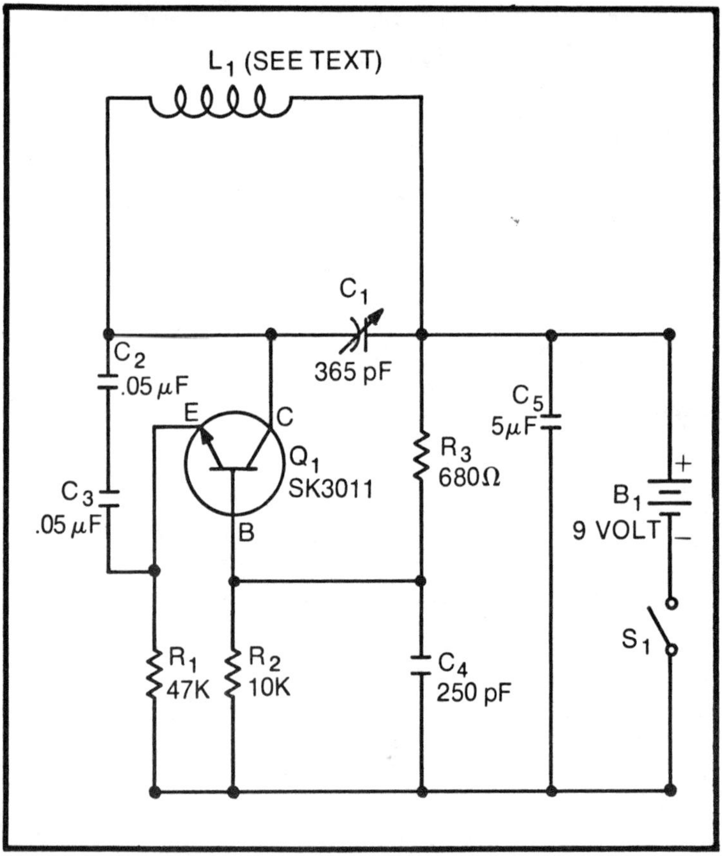

Fig. 3-2. Schematic wiring diagram of the metal detector circuit described in this chapter.

Have these two drawings in front of you the whole time you are assembling the circuit components.

Winding The Coil

Since the construction of the coil and its housing is going to require some waiting periods for the various applications of cement and varnish to dry, you should begin the winding of the coil and the construction of its housing (or form) first.

Temporarily secure one end of the 22-gauge magnet wire to a cardboard winding form—leaving approximately 8 inches of lead wire hanging beyond the tape. Then, scamble-wind 18 turns of the magnet wire around the form, securing the entire coil—after the final turn—with another piece of tape. Reel out another 8 inches of the wire beyond this tape and cut with your diagonal pliers.

Be careful to keep the coil intact while you remove it from the cardboard form; collapse the form if necessary. Temporarily wrap a few pieces of masking tape around the coil to help hold the coil in shape while you test it. Later you will want to saturate the wires with coil dope to hold the strands firmly in place, because one unstable wire will cause the operation of the entire unit to be unstable also. The search coil at this stage should look like the one in Fig. 3-3.

Collect the three paper plates together, spread a thin coat of epoxy cement on the bottom of one plate (Fig. 3-4), press another plate on top of this, and repeat this same procedure. Once together, press the plate surfaces firmly around the outside edge as shown in Fig. 3-5. You should now have three paper plates firmly secured to each other as shown in Fig. 3-6. When the epoxy is dry (in about 6 minutes) varnish the entire assembly. Depending upon the type of varnish you use, the drying time could take from two to four hours, so make sure this is one of the first steps you undertake for this model of metal detector. While the varnish is drying, you can begin

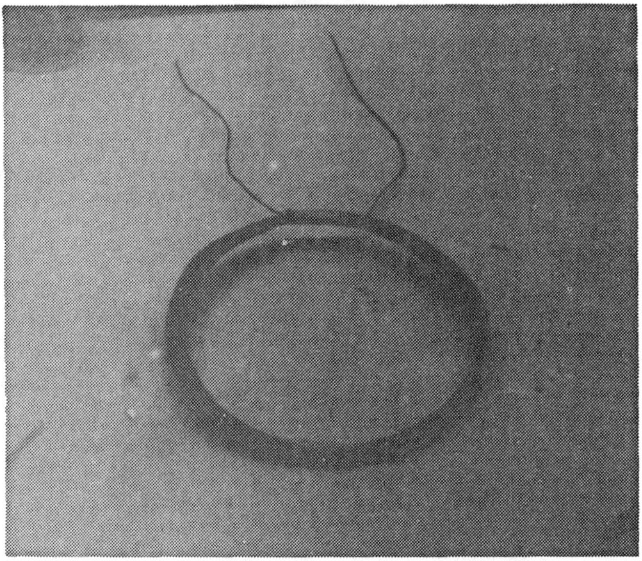

Fig. 3-3. Appearance of circuit coil after being removed from the cardboard form.

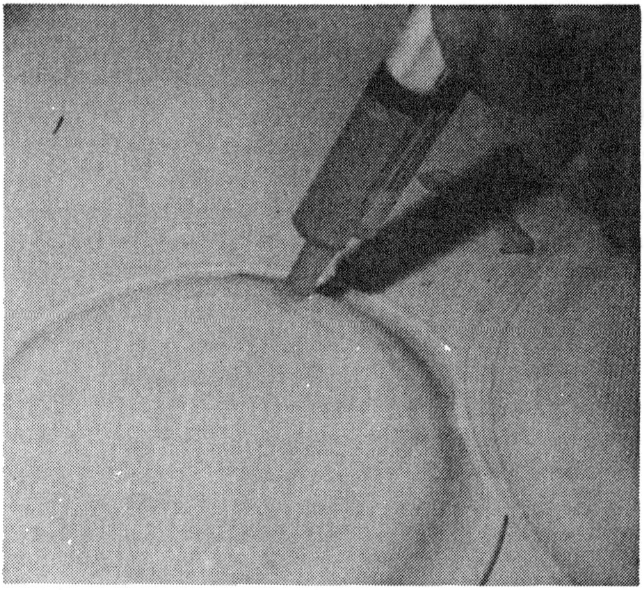

Fig. 3-4. Method of cementing paper plates together with epoxy.

assembling the components on the circuit board according to the information in Figs. 3-1 and 3-2.

Wiring The Circuit Board

As mentioned previously, place both the pictorial and wiring diagram in a position near you where you will be able to

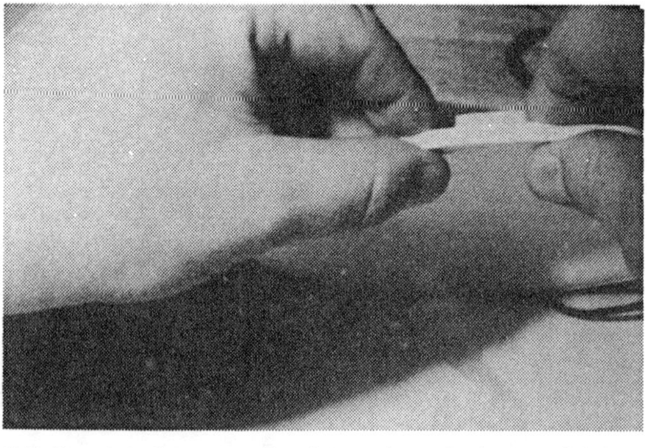

Fig. 3-5. Once together, press the plate surfaces firmly around the outside edge as shown here.

refer to both of them at a glance. Once you have made a temporary connecton by twisting the leads together, double-check the wiring diagrams to make certain that they are correct before soldering the joints.

First, lay all of the components needed within easy reach, as well as all necessary tools and other materials. Prior to the actual assembly, your work area should look something like the photo in Fig. 3-7. One way of assembling the components is as follows:

1. Enlarge two holes in the circuit board as shown in Fig. 3-1, and mount the battery holder with two 2-56 × 1/2-inch machine screws and two 2-56 hex nuts.
2. Apply epoxy cement to the bottom of the trimmer (variable) capacitor, and secure to the opposite end of the circuit board from the battery holder. You may want to secure this trimmer and circuit board in a small vise while the epoxy is drying for about ten minutes; or place some weight on the trimmer to ensure a tight joint while the epoxy is drying.
3. Insert the three lead wires of the transistor in three appropriate holes in the circuit board in the

Fig. 3-6. Three paper plates firmly secured to each other to be used as a search head to house the search coil.

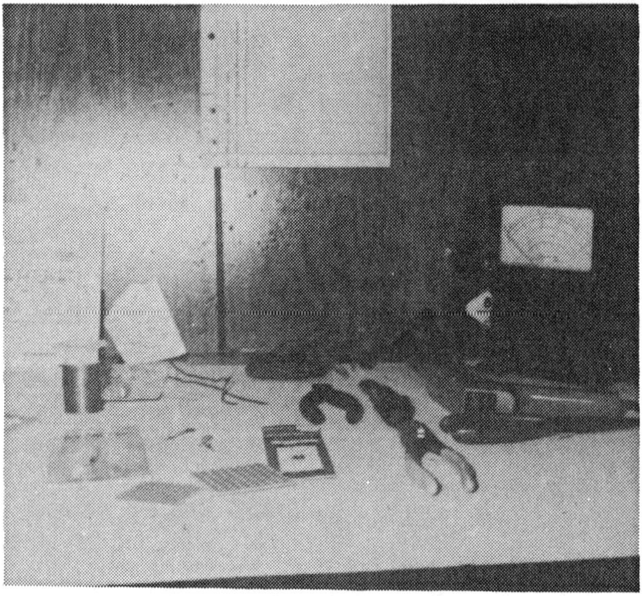

Fig. 3-7. Your work area should look something like the one shown here before you begin assemblying the circuit board.

approximate location as shown in the pictorial diagram in Fig. 3-1. Next insert one of the leads from resistor R1 in a hole near the emitter lead of the transistor, and wrap the two leads together to hold the components in place until you solder them later.

4. Insert capacitor C3 and twist one lead around the leads jointed earlier; that is, one lead from resistor R1 and the emitter lead from the transistor. Continue by inserting capacitor C2 near capacitor C3 and twist the remaining unused lead from capacitor C3 around one of the leads from capacitor C2. The remaining lead from capacitor C2 is then wrapped around the collector lead from the transistor.
5. Cut a piece of hook-up wire long enough to reach from the last joint made to one of the terminals of the trimmer C1. Go ahead and solder the wire to the terminal at this time, but leave the other end of the wire twisted around the leads only.
6. Return to the unused lead from resistor R1. Twist around this lead: one of the lead wires from resistor R2; one of the lead wires from capacitor C4; one of the

lead wires from capacitor C5; and one piece of hook-up wire long enough to reach from this joint to the location of one of the switch terminals. In case you want to omit the single-pole switch, attach one of the battery clip leads to this joint in place of the piece of hook-up wire; just make certain that the wire will reach from the joint in question to the location of the battery holder.

7. The remaining wires (the ends unused) from R2 and C4 are wrapped around the base lead on the transistor, as well as one lead from R3.
8. Wrap the unused lead wires from R3 and C5 together as well as the positive battery clip wire. Cut a piece of hook-up wire long enough to reach from this joint to the unused terminal of the trimmer (variable) capacitor C1, and temporarily wrap both ends until they can be soldered.
9. You have just roughed-in the circuit components on the circuit board, but before you clip the excess lead wires and solder, check your work against the wiring

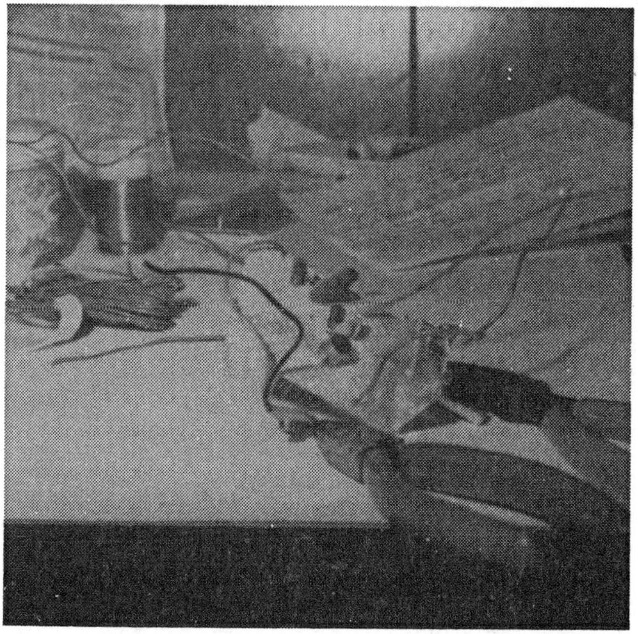

Fig. 3-8. Top view of circuit board showing components, hook-up wire, and battery clip leads.

Fig. 3-9. Bottom view of circuit board prior to soldering the connections and clipping the leads.

diagrams in Figs. 3-1 and 3-2. When you are certain that your work is identical to these illustrations, and looks like Fig. 3-9, clip the excess leads and solder according to instructions given in Chapter 1.

Now that you have finished the electronic work on the circuit board, we can now get back to the search coil and various mechanical items required for completing the detector.

10. The varnish should now be dry on the paper-plate coil form, so place the coil on the inside of the plate, mark off the location of two holes for the coil leads to go through, and puncture at these two marks with an ice pick or similar item.
11. Insert the two leads through their proper holes, apply epoxy cement to the side of the coil which is next to the paper plate, and press the coil firmly into position. A medium-heavy weight should be placed on top of this coil to hold it firmly against the plate until the epoxy is dry.

12. The epoxy, holding the coil, should be dry in about 10 minutes. Then, turn the plate over, apply some epoxy in the center of the plate, and place the wooden handle on top of this spot to secure the plates to the handle. However, before cementing the handle to the plates, cut the end of the wooden handle at an angle of approximately 30° so that the plate will move parallel to the ground when the detector is in the operating position.
13. After the epoxy holding the plates to the handle becomes dry, secure the circuit board to the handle with a wood screw. See Fig. 3-10. The lower end of the circuit board should be approximately three inches up from the plate.
14. Connect the two coil leads (protruding through the two holes in the paper plates) to capacitor C1—one coil lead to one terminal, and the other lead to the remaining terminal of C1.
15. Place a small portable transistor radio about 30 inches up on the handle from the circuit board, and secure it firmly to the handle with plastic electrical tape; two wraps around each end of the radio will suffice. You are now just about ready to try out your project.

Fig. 3-10. View of search head, wooden handle, and search coil leads attached to the circuit board prior to mounting the latter on the wooden handle.

Fig. 3-11. Metal detector described in this chapter being used to hunt for metallic objects.

TESTING

Insert a 9-volt transistor radio battery into the battery holder and connect the battery clip. If you have omitted the single-pole switch, this completes the connection on this end of the detector. Next turn on the transistor radio and tune it to a weak station. Then adjust the trimmer (C1) until you hear a beat-note (whistle). These adjustments should be done outside, away from all metals.

Adjust the volume and pitch of the tone to a comfortable setting before sweeping the metal detector over some metal object. As metal is detected, the pitch of the tone will change either up or down—usually up.

With a little playing around, you should be able to adjust the detector so that no whistle is heard except when metal is brought near the search coil. This eliminates that boring constant whistle in your ear the whole time you are searching. To do this, adjust the trimmer (C1) until a whistle is heard; then readjust slightly for a zero-beat frequency—that point of quiet between whistles. When the search coil passes over metallic objects of sufficient size, the whistle will return. If swinging the search coil in a sweeping motion, you will probably hear only a short "beep" as the coil passes over the object. When this is heard, stop walking, and move the search coil slowly over the approximate area in which you heard the "beep" in order to pinpoint the exact location of the "find".

OPERATION

You will obtain the best results from this instrument when searching over smooth surfaces such as at a beach, or on closely mowed grass. When testing this detector (Fig. 3-11), it was found that blades of grass hitting the coil as it was being swung back and forth, caused a false "find" signal. You will also find that the more you experiment with this model, the smaller the objects you will be able to find. On the first couple of tests, metallic objects about the size of a tin can top were easily detected up to 3 inches deep; coins could be detected at depths of about 2 inches (although this depth can be improved with practice) if the user listens very carefully for the tone change.

The accompanying illustration (Fig. 3-12) shows some of the "finds" possible with this inexpensive detector; they

Fig. 3-12. Various items detected with this locator include: an axe head, a .36 caliber muzzle loading rifle barrel, a Civil War artillery shell, a steel trap, and several lead bullets.

include: an axe head, a .36 caliber muzzle-loading rifle barrel, a steel trap, a fragment of an artillery projectile and several Minie balls—all found on the Civil War battle ground where the battle of Milford, Va. was fought (the village is now called Overall, Va.).

You will note from the illustration of this detector, that no housing was constructed for the circuit board. During fair weather and over smooth terrain, a housing is not necessary—the circuit board is merely screwed to the wooden handle. However, you may prefer to use a small plastic box to house the board. Since the capacitor C1 is cemented to the board itself, this capacitor's dial shaft will hold the board firmly in place inside the housing once a locknut is installed on the threads of the shaft mounting sleeve. Those plastic boxes that many electronic kits are sold in will work fine.

Chapter 4

Self-Contained Detector

The circuitry in this type of detector is typical of those commercial models that sell from between $30 to $50 off the shelf. You can build the same quality detector—following directions in this chapter—for less than $20! Besides having a complete self-contained unit (requiring no other oscillator like the transistor radio in the first two examples), this unit also boasts an external speaker and a meter to further aid you in your search for "treasures".

While not too much more difficult than the first circuits given earlier in this book, the construction of this detector is more time consuming. Therefore, if this is one of your first electronic projects, we recommend that you take two evenings to complete the project rather than try to complete it in one session. You won't get nearly as tired by doing so, and your chances of becoming discouraged are less.

A pictorial diagram of this project is not included, because if you have completed the two preceding circuits, you should be able to design your own physical layout of the components by this time.

PARTS

Since there are more components involved in this project than the previous two, you should be extremely careful to follow the instructions given in the text, as well as performing

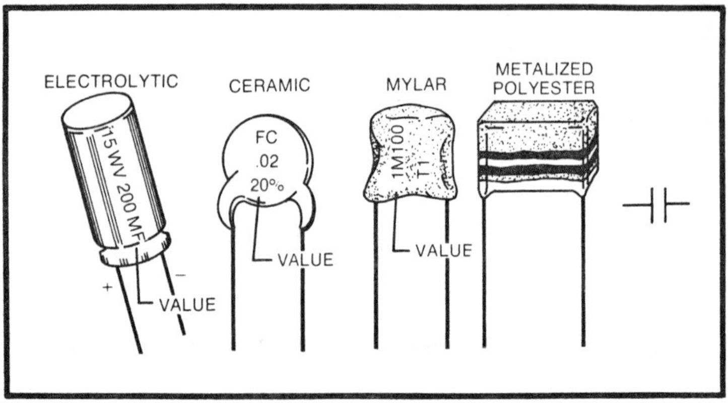

Fig. 4-1. Four different types of capacitors with their schematic symbol.

each step in the order given. Before beginning the actual construction, let's review some of the components which are required for this project.

Capacitors

You can see from the drawings in Fig. 4-1 that capacitors come in many different shapes and sizes. Most of them have the capacitance value stamped on the case as well as other information such as percentage tolerance, heat tolerance, maximum voltage, and the like. Only the value, however, is given in the parts lists for the circuits in this book. Note that the electrolytic capacitor in Fig. 4-1 has minus and plus leads indicated; these leads must be connected properly (the correct polarity) in order for this type of capacitor to function correctly.

Transistors

Depending upon the metal detector circuit you are building, you will use one or more of the four types of transistors shown in Fig. 4-2. It is extremely important to connect the right lead to the correct location, as each lead has a different function and the device will not work properly if improperly connected.

Diodes

The metal detector circuit in this chapter requires the use of one diode to detect a signal for the meter. It is important that it be wired in the proper direction; that is, the cathode and

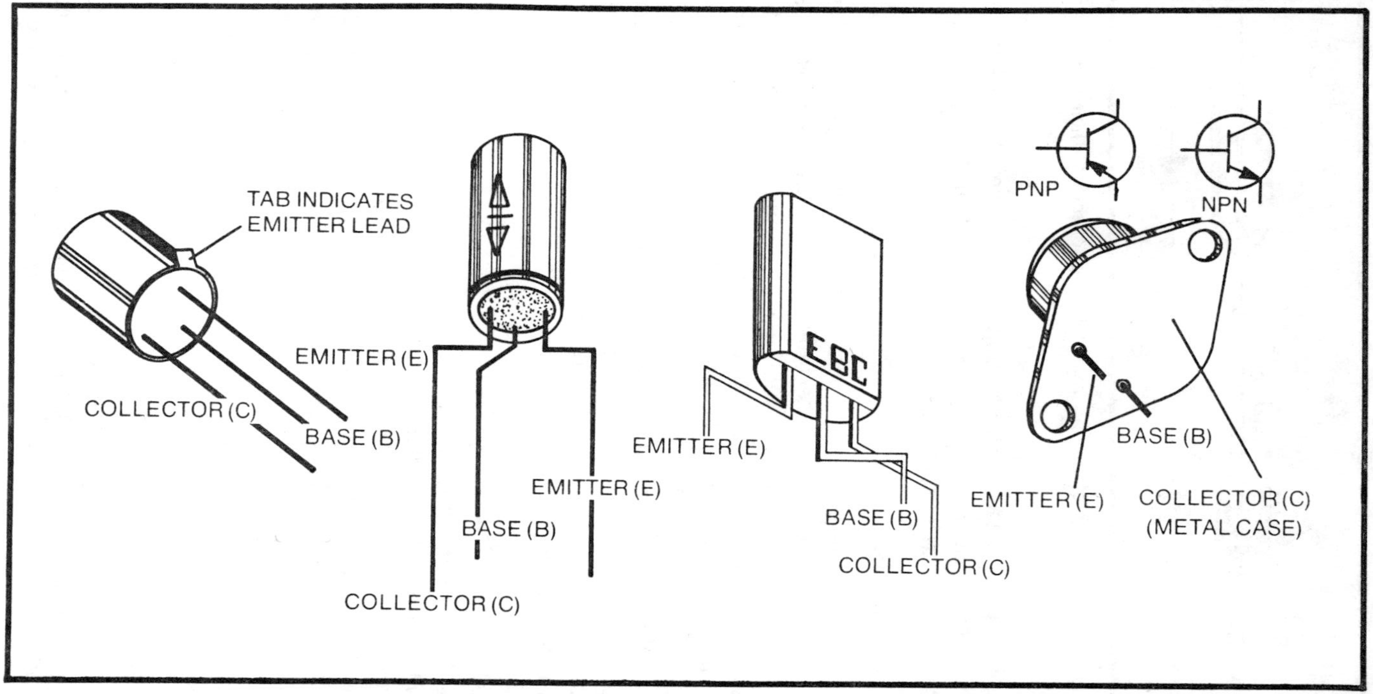

Fig 4-2. Four different types of transistors with the svmbols for PNP and NPN transistors.

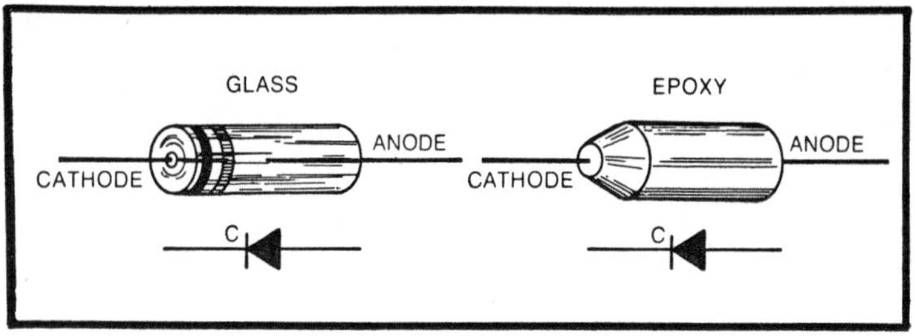

Fig. 4-3. Two types of diodes shown in relationship to their schematic symbol.

anode ends arranged according to the schematic diagram. Figure 4-3 shows two types, each of which is in proper relationship to its schematic symbol.

Complete List

Check the schematic diagram of this metal detector circuit in Fig. 4-4 to become familiar with the various components. Besides conventional electronic tools, resin core solder, epoxy, and various mechanical parts to construct the handle and housings, you will need the electronic components listed next to Fig. 4-4.

You will also need an assortment of hexnuts and machine screws and such miscellaneous items as hook-up wire, epoxy, coil dope, etc. Accumulate all of the necessary parts before you begin construction of the unit.

ASSEMBLY INSTRUCTIONS

You will need a cardboard, plastic, or wooden form approximately 5 inches in diameter on which to wind the search coil (L1). Follow procedures given earlier, and scramble-wind 13 turns of No. 22 AWG magnet wire around the form, leaving 6 to 8 inches of lead wire on each end. Go ahead and also construct the handle, search head, and other necessary mechanical items at this time. Saturate the coil in coil dope, and while this is drying, begin the construction of the circuit board.

The Circuit Board

Place the wiring diagram in a position where it can be easily seen at all times, and have all the electronic components laid out in an orderly manner, and within easy reach. After you have studied the schematic diagram (Fig. 4-4) thoroughly, roughly install the various components on your circuit boards. Do not make any connections at this time. Study your first layout, and then go back and rearrange the components—if necessary—so that they are arranged neatly and also in a position so that they can be wired easily.

Remember that the 3 potentiometers, the switch, the speaker, and the meter will be mounted on the aluminum chassis—only the remaining circuit components will be mounted directly on the circuit boards.

Examine coil L2, and notice that the bottom of the unit has two holding clips and five pins—three pins in one row and two pins in the other row. Place this unit so that the side with the row of two pins is on your right as you look at the can. Clip both of the pins off in the two-pin row and the first pin (the one closest to you) in the three-pin row. You now have only two pins left on the bottom, Insert these two pins through holes in the circuit board as well as the two holding clips so that the pins are on the bottom of the circuit board; you may have to drill a couple of new holes to make this fit. Bend the holding clips so that the unit is secured to the circuit board.

Bend the leads of the four transistors approximately as shown in Fig. 4-5, and insert these leads through the appropriate holes in the circuit board. Make certain that you have properly identified the leads and have them arranged so that other components may be connected to them. A wrong connection can damage the transistor.

Select a good location for the battery and its holder and secure the battery holder to the circuit board with the two screws and two nuts usually provided with the holder.

Locate and connect the remaining components by following procedures given in Chapter 1. However, do not solder the connections until you have checked them to make certain that you haven't made a mistake. Once you are sure of you connections, solder and clip the excess lead wires from each connection.

Final Assembly

When completed, securely mount the circuit board inside of the aluminum housing—making certain that the bottom of the circuit board does not touch the metal portion of the housing. It is assumed that you have already mounted the

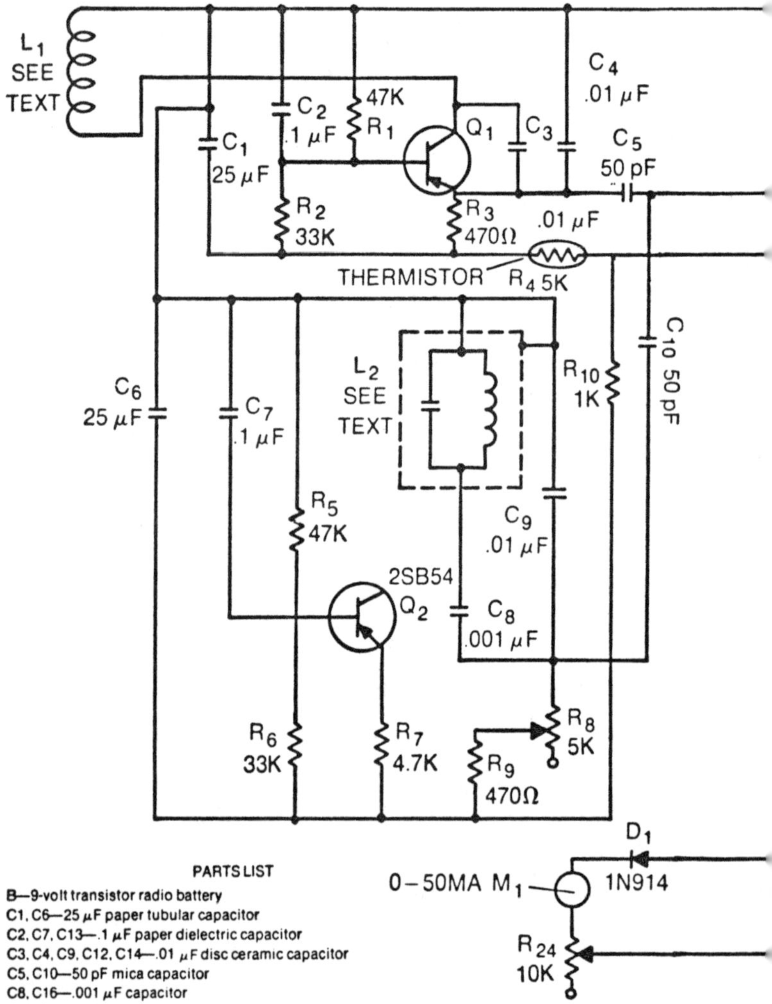

PARTS LIST

B—9-volt transistor radio battery
C1, C6—25 μF paper tubular capacitor
C2, C7, C13—.1 μF paper dielectric capacitor
C3, C4, C9, C12, C14—.01 μF disc ceramic capacitor
C5, C10—50 pF mica capacitor
C8, C16—.001 μF capacitor
C11, C15—5 μF electrolytic capacitor
C17—180 pF mica capacitor
D1—1N914 diode (Radio Shack 276-1144 or equal)
IC—PA234 (GE) integrated circuit
L1—13 turns No. 22 AWG enameled magnet wire scramble wound on 5-inch diameter form
L2—adjustable coil (Radio Shack 99-301 or equal)
M1—0–50 mA meter
Q1, Q2, Q3, Q4—2SB54 PNP transistor (Radio Shack 276-2007 or equal)
R1, R5—47K carbon resistor
R2, R6—33K carbon resistor

Fig. 4-4. Schematic diagram of the metal detector circut described in this chapter.

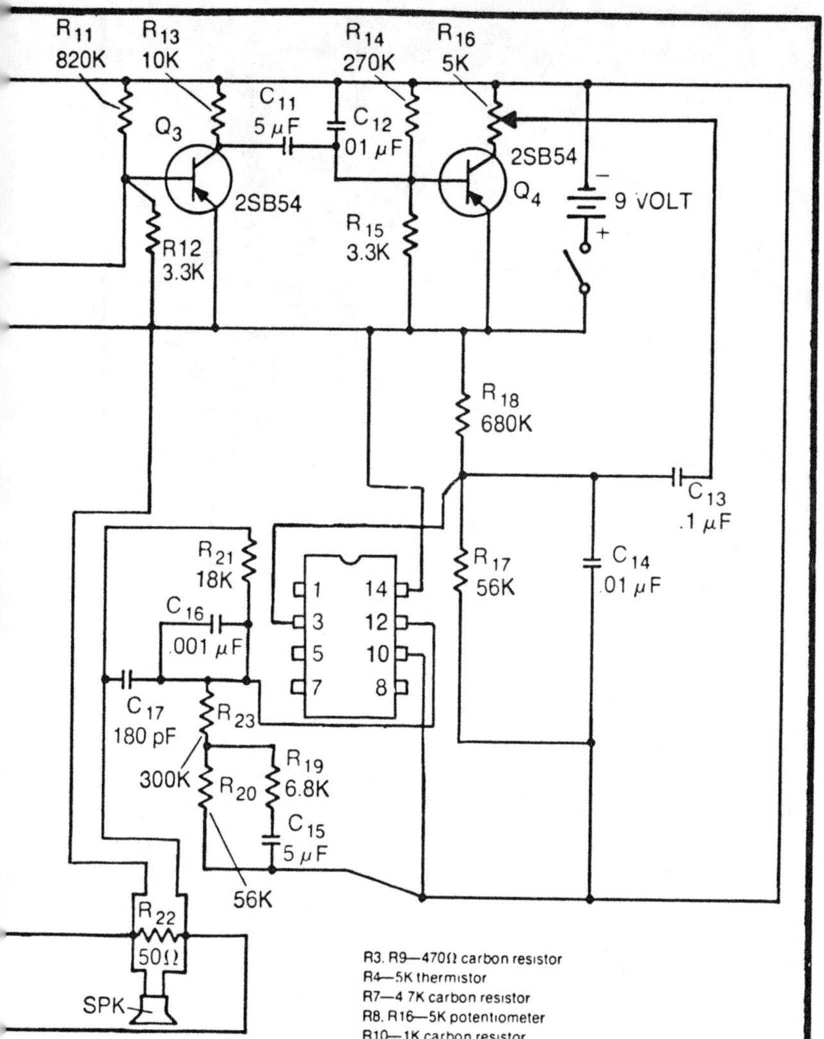

R3. R9—470Ω carbon resistor
R4—5K thermistor
R7—4 7K carbon resistor
R8. R16—5K potentiometer
R10—1K carbon resistor
R11—820K carbon resistor
R12. R15—3 3K carbon resistor
R13—10K carbon resistor
R14—270K carbon resistor
R17. R20—56K carbon resistor
R18—680K carbon resistor
R19—6.8K carbon resistor
R21—18K carbon resistor
R22—50Ω carbon resistor
R23—300K carbon resistor
R24—10K potentiometer
All resistors 1/2 watt. 20% tolerance
Misc—21 ft No 22 AWG enameled magnet wire. 4 ft two-wire shielded cable. 2 shielded cable connectors. 1 aluminum box to house electronic components

Fig. 4-5. Method of bending the leads for transistors used in this circuit.

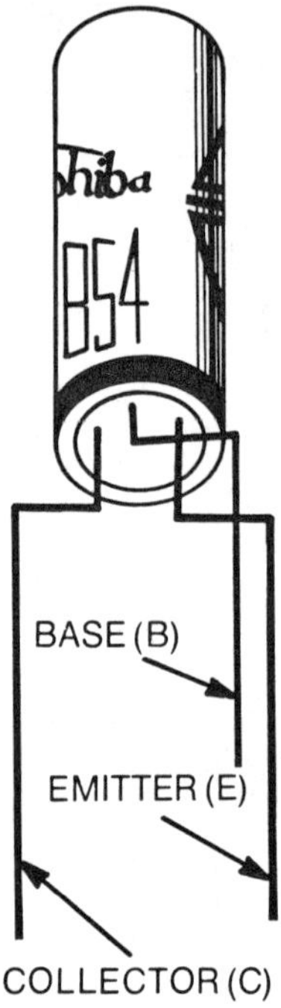

meter, switch, speaker, and potentiometers to the housing by cutting out the proper openings through the walls of the housing, and then securing the various components with either locknuts or self-tapping screws. Once the board is in place, finish connecting the meter, switch, etc., to the appropriate places on the circuit board.

Your next step is to connect the two leads from the search coil to the circuit board by means of the shielded cable and connectors. Twist this cable around the handle as it climbs the handle from the search head; cut off any excess.

TESTING

Take the unit outside away from all metallic objects, and flip the switch to the "on" position; leave the housing cover off at this time, so that any adjustments that may be necessary can be made. Turn R16 (volume control) fully clockwise and potentiometers R8 and R24 to approximately mid-rotation. Insert a small screwdriver in the slot on top of L2, and tune (either clockwise or counterclockwise) until an audible tone is heard. Continue turning the slot until the pitch lowers and dies out completely; stop the instant the pitch is gone.

Now adjust the three potentiometers until:

1. A comfortable hearing level is obtained.
2. The meter needle is approximately 1/4-inch up from the "0" mark on the dial.

When the tuning is completed, you can search for metal objects in either the ground or in walls by slowly sweeping the search head parallel to the surface in which you are searching. As metal is detected, the pitch will change to a higher pitch (for most metals) and the meter needle will deflect upward—probably to full scale.

Chapter 5
Miscellaneous Detector Circuits

The metal detector circuits in Chapters 2, 3, and 4 were designed to acquaint the reader with the basic circuitry of beat-frequency type detectors. All of them are easy to build with standard replacement parts available at local distributors. The additional circuits covered in this chapter, while somewhat more advanced, are similar to those previously covered, that is, easy to build in a short period of time and if you don't have enough spare parts laying around your bench, they are easily obtainable at your local electronics shop.

Besides the enjoyment and satisfaction you'll derive from building your own metal detectors, you will also save money—lots of money if you were to purchase the same detector off the shelf. Therefore, since saving money is certainly one of the reasons for building your own metal detector, you should try to use as many spare parts that you have around your home as possible.

SUBSTITUTE COMPONENTS

So you can't find some of the exact components as specified in the parts list—what do you do then? Substitute! That's right, with a little knowledge of electrical values, you'll be able to substitute what parts you do have for those that are specified in the component lists accompanying each detector circuit drawing. For example, if a component required for one of the circuits in this book happens to be a 0.05 μF disc capacitor, but all that you have on hand are capacitors rated at 0.1, merely connect two of the 0.1 μF capacitors in series; the new value will be:

$$\frac{.1 \times .1}{.1 + .1} = .05\ \mu F$$

On the other hand, if you wanted to use two .05 μF capacitors in place of one capacitor rated at .10 μF, connect the two .05 μF capacitors in parallel. Therefore, when connecting two capacitors of the same value in series, their capacitance will be reduced to half the value of one of the capacitors. If two capacitors were connected in parallel, the value of one of the capacitors would double.

Resistors are calculated exactly opposite from capacitors; that is, resistors connected in series will have a value equal to the sum of all the resistors. When connected in parallel, the resistance will decrease. Before continuing, let's take a closer look at these calculations. Your knowledge of them will be tremendously helpful when substituting components in electronic circuits.

The total capacitance of a parallel circuit is equal to the sum of the separate capacitors.

Total Capacitance = Capacitor 1 + Capacitor 2 + Capacitor 3, etc.

For example, if two capacitors are connected in parallel as shown in Fig. 5-1, and each has a value of 4.7 μF, the total capacitance of the circuit will be 4.7 + 4.7 = 9.4 μF. If more than two capacitors were connected to this same circuit, the total capacitance of the circuit can be found by merely adding the remaining values.

The rule most often used to find the total capacitance in a series circuit is:

> The sum of the reciprocals of the separate capacitors equals the reciprocal of the equivalent capacitance.

Since the reciprocal of any number is 1 divided by the number, the formula for finding the total capacitance of two or more capacitors connected in series is:

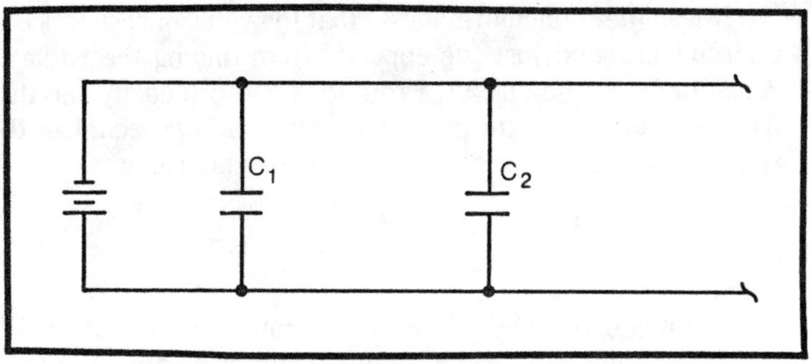

Fig. 5-1. Schematic diagram showing two capacitors connected in parallel.

$$\frac{1}{C_t} = \frac{1}{C_1} + \frac{1}{C_2} + \frac{1}{C_3} + \text{etc.}$$

To illustrate, let's find the total capacitance of the capacitors shown in the circuit in Fig. 5-2. Substituting in the formula, we have,

$$1/C_t = 1/.30\ (\mu F) + 1/.5\ (\mu F) + 1/.1\ (\mu F)$$

or:

$$\begin{aligned} 1/C_t &= .333 + 2 + 10 \\ 1/C_t &= 12.333 \\ C_t &= 1/12.333 \\ C_t &= .08\ \mu F \end{aligned}$$

The capacitance of the circuit in Fig. 5-2 is 0.08 μF. The total capacitance of capacitors in series is always less than the value of the smallest capacitor in the series, and the same is true for resistors in parallel. The resistance of the parallel circuit is less than the value of the smallest resistor in it.

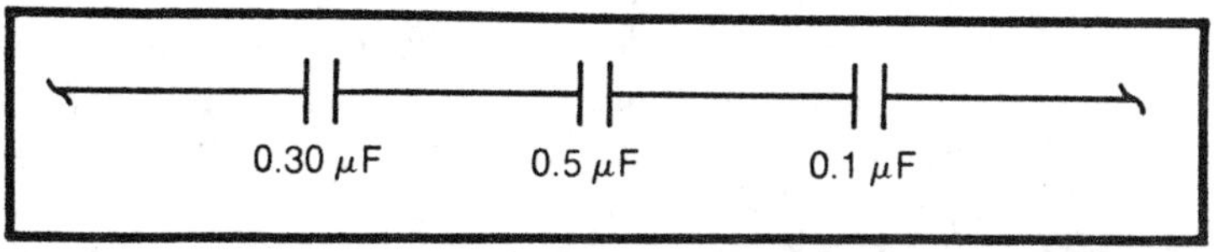

Fig. 5-2. Schematic diagram showing three capacitors connected in series.

It was mentioned previously that the value of resistance in a circuit is found just the opposite from finding the value of capacitance in the circuit. From this, you can easily see that the total resistance in a series circuit would be equal to the sum of all resistors in the circuit—the formula being,

$$R_t = R_1 + R_2 + R_3 + \text{etc.}$$

When substituting "C" for capacitance in place of "R" in this same formula, we have a means of finding the total capacitance in a parallel circuit; that is

$$C_t = C_1 + C_2 + C_3 + \text{etc.}$$

Likewise, the formula given previously for finding the total capacitance in a series circuit, may be used to find the total resistance in a parallel circuit:

$$\frac{1}{R_t} = \frac{1}{R_1} + \frac{1}{R_2} + \frac{1}{R_3} + \text{etc.}$$

Remember also that nearly every component value given in these circuits can vary as much as 20% without losing too much detector sensitivity—so begin using up those spare parts!

While constructing any of the metal detector circuits to follow, pay close attention to the suggestions given in Chapter 1, as well as the chapters following. Hardware may be the same as described for the detectors in Chapters 1 and 2: paper plates or a plastic Frisbee to house the search coil; broom handle, wooden dowel, or aluminum tubing for the handle; aluminum or plastic box to house the circuit components; and conventional screws, bolts, and epoxy to secure the components.

Regardless of the type detector you build, you will have to practice considerably in order to get the most use out of it. In fact, it has been said that more "finds" can be had by using the poorest metal detector—with which you are familiar—than with the best metal detector if you're not familiar with its operation. Chapter 6 offers suggestions on how to adjust your detector as well as the correct use of it.

ANOTHER DETECTOR USING A PORTABLE RADIO

The metal detector circuit illustrated in Fig. 5-3—while somewhat more sophisticated than those described previously—is to be used in conjunction with a small portable transistor radio.

Begin construction of this metal detector by winding the coil L1 on a cardboard or plastic form allowing approximately 10 inches of extra lead wire on each end. Coil L1 consists of 14 turns of No. 18 AWG enameled magnet wire scramble-wound on a 12-inch diameter form. Saturate the wound coil with coil dope or GE adhesive, and continue on to the circuit board.

The arrangement of the electronic components (listed in Fig. 5-3) on the circuit board will be similar to the arrangement of the circuit described in Chapters 2 and 3; that is, the trimmer may be mounted near the edge of the board so that the trimmer's shaft can be used to support the entire board once the trimmer is cemented to the board. The remaining components are connected using the procedures outlined in Chapter 1, as well as other places in the book.

The circuit board is then housed in some convenient-size metal box like the one specified for the economy detector in Chapter 2. A Frisbee or a paper plate may be used for holding the search coil. Although any type of two-wire insulated cable may be used to connect the circuit board to the search head coil, shielded cable is recommended for this and all other detectors listed in this chapter.

METAL DETECTOR METER CIRCUIT

For those of you who want to add an indicating meter to any of the circuits in this book, the circuit in Fig. 5-4 provides a simple means of doing so.

The number of turns in the coil (L1) will vary depending upon the type of metal detector used; however, one or two turns of No. 18 AWG enameled magnet wire attached to the search head should be sufficient for most applications.

If your detector circuit board has enough room, the components may be mounted directly on this board, and the meter may be then mounted in the housing in which the detector circuit board is mounted. The coil (L1) mounted on the search head may then be connected to the electronic components by means of any two-wire shielded cable—much the same as the search coil itself.

PARTS LIST

B—9-volt transistor radio battery

C1—365 pF variable capacitor

C2—1000 pF mica capacitor

C3—680 pF mica capacitor

C4—.02 μF paper tubular capacitor

D1—1N914 diode (Radio Shack 276-1144 or equal)

L1—14 turns No. 18 AWG enameled magnet wire scramble wound on a 12-inch diameter form

Q1—3N128 transistor

RFC—2.5 mH RF choke

R1—22K 1/2 watt resistor

S1—single-pole, single-throw switch

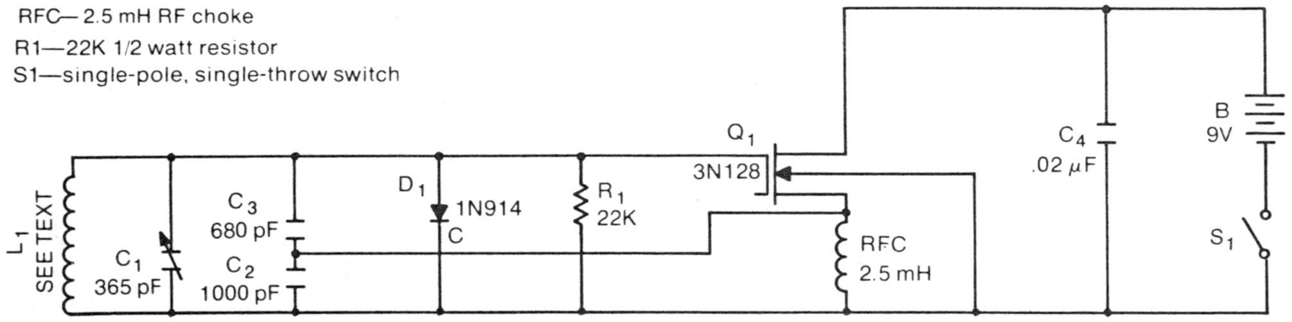

Fig. 5-3. Metal detector circuit to be used in conjunction with a small portable transistor radio.

When all components are in place and all connections are made, adjust R3 until the pointer on the meter reads mid-scale. When the search head comes into close proximity with metallic objects, the pointer (needle) will deflect either up or down, depending upon the type and size of the metal object.

ANOTHER SELF-CONTAINED DETECTOR

A self-contained metal detector such as the one shown in Fig. 5-5 can be built in only a couple of hours and with a total

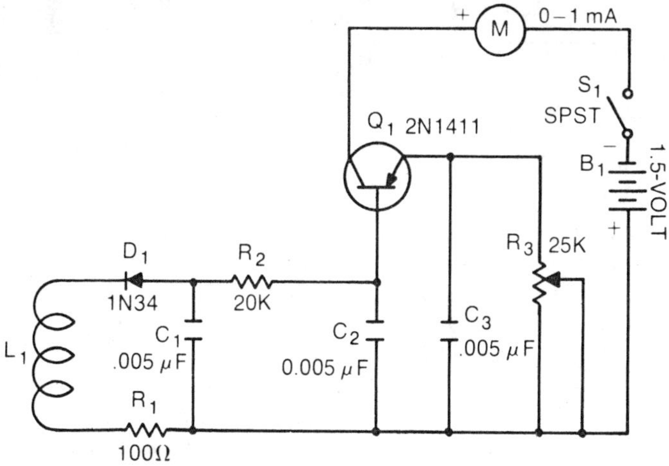

PARTS LIST

B—1.5-volt penlight battery
C1, C2, C3—.005 μF disc ceramic capacitor
D1—1N34 diode (Radio Shack 276-1102 or equal)
L1—one or two turns of No. 18 AWG enameled magnet wire attached to search coil form
M—0–1 mA meter (Olson Electronics ME-215 or equal)
Q1—2N1411 transistor (Radio Shack 276-2003 or equal)
R1—100 ohm 1/2 watt carbon resistor
R2—20K 1/2 watt carbon resistor
R3—25K linear potentiometer
S1—single-pole, single-throw switch

Fig. 5-4. Simple metal locator circuit utilizing a 0–50 mA meter to indicate the presence of metal objects.

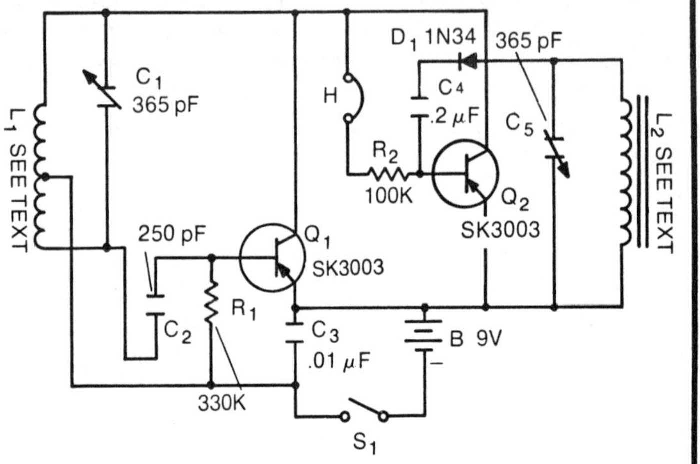

PARTS LIST

B—9-volt transistor radio battery
C1, C5—365 pF variable capacitor
C2—250 pF mica capacitor
C3—.01 μF apper tubular capacitor
C4—.2 μF paper tubular capacitor
D1—1N34 diode (Radio Shack 276-1144 or equal)
H—high impedance magnetic head phones
L1—20 turns No. 22 AWG enameled magnet wire scramble wound on a 6-inch form, tapped at 10 turns
L2—ferrite rod antenna coil (Radio Shack 270-1430 or equal)
Q1, Q2—SK3003 transistor (Radio Shack 276-2004 or equal)
R1—330K, 1 2 watt carbon resistor
R2—100K, 1 2 watt carbon resistor
S1—single-pole, single-throw switch

Fig. 5-5. Self-contained metal detector circuit using headphones to amplify the beat whistle when metal objects are found.

cost of components not exceeding $20—including the head phones. Such a detector will be as sensitive as any of the circuits designed for use in conjunction with a small portable transistor radio. You may even find that this model is slightly more sensitive due to addition of the head phones, which will allow you to detect some of the faint signals of deeper metallic objects that you otherwise would miss.

When purchasing the materials and components for this detector, don't overlook the possibility of using material which you already have on hand. Remember that we discussed the possibility of substituting components—using methods given earlier in the chapter—and that components may vary in their rating without any adverse affects.

Again, begin the construction of this detector by winding the coil on a 6-inch form, applying coil dope, and letting this dry while you construct the mechanical parts and construct the circuit boards. The coil consists of 20 turns of No. 22 AWG enameled magnet wire and is tapped at 10 turns. The circuit parts are then connected as described earlier in this book.

One word of caution. When using head phones on this detector—or any other project for that matter—always tune the set prior to placing them over your ears. The volume of a strong signal confined within the boundaries of your ears could cause damage to your hearing.

A SENSITIVE DETECTOR

This particular metal detector circuit uses a center tap on the search coil and its small 4-inch diameter makes it extremely sensitive to small objects such as coins, rings, and Minie balls. It is a complete unit and requires only a set of head phones to indicate a find. However, the meter circuit described previously may be used in conjunction with the head phones or substituted in place of them.

Besides the mechanical items like the handle, housing, and search head, you will need the electronic components listed in Fig. 5-6.

This particular metal detector will enable you to really get in on the fun and excitement of treasure hunting. You will be able to find coins, antiques, buried metal objects of any type—fast and easy. Besides being fun to use, you will also find that they're practical when used to locate buried plumbing pipes, metal cables, and the like; and don't overlook the possibilities of locating hidden electrical wires in partitions also—ones that you may be cutting into on your next home improvement project.

The entire detector can be assembled in a couple of evening sessions and the entire list of parts should cost no

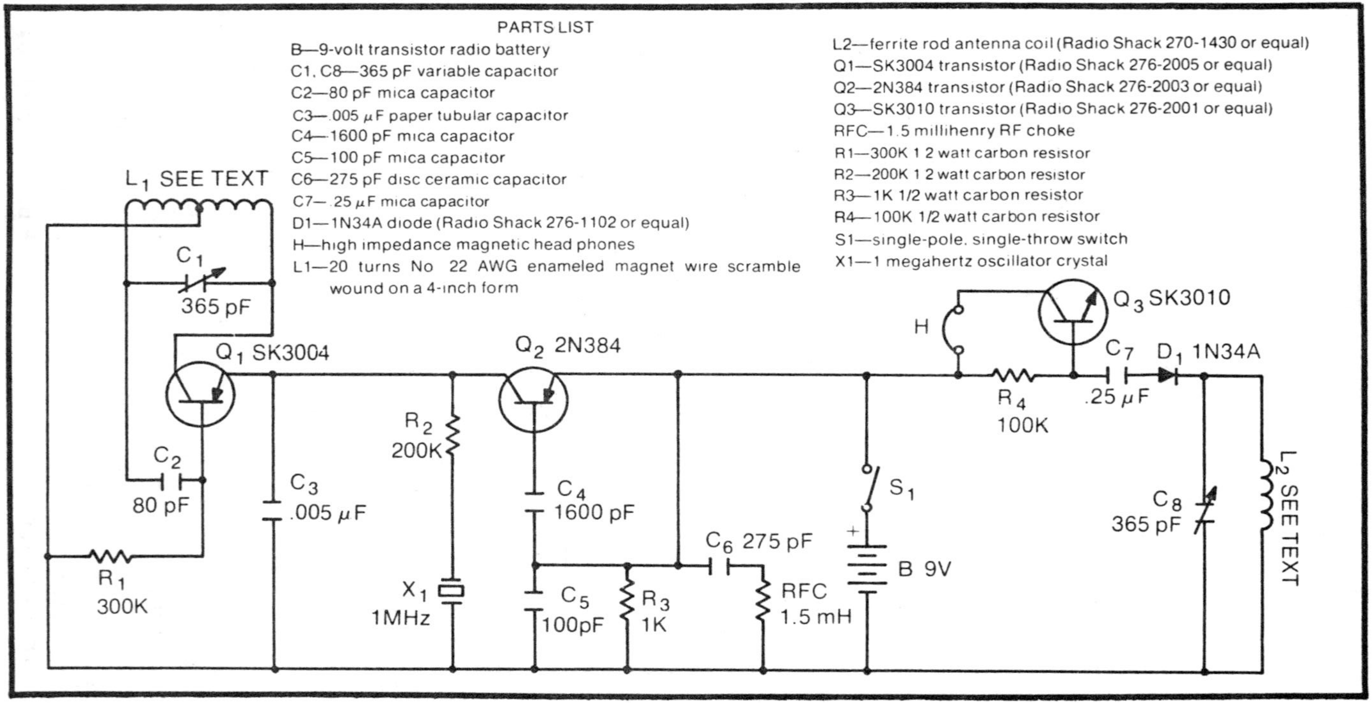

Fig. 5-6. Sophisticated beat frequency oscillator type metal locator utilizing headphones to amplify the "beat" tone.

more than $15 to $20. If you have some of the parts laying around your home or shop, so much the better. If you don't have exactly what is specified, substitute some components of different value by connecting two or more in series or parallel—there are all kinds of possibilities.

When constructing the circuit board, make sure that you keep the circuit schematic diagram shown in Fig. 5-6 in view at all times in order for you to check your connections while making them and especially before soldering. Always double-check your work before soldering—this could save a lot of time if you have to do the work over due to an error on your part.

The coil (L1) for this detector consists of 20 turns of No. 22 AWG enameled magnet wire wound on a 4-inch coil. Tap the center of the coil (10th turn) as shown in the schematic diagram in Fig. 5-6. A cardboard Quaker Oats box is exactly the correct size for this coil.

ANOTHER DETECTOR USING A METER

Another metal detector circuit utilizes a 0–50 mA meter to indicate a "find". Once the handle, housing, and search head is constructed, you should be able to complete the balance of this project in less than an hour—then you're on your way to finding hidden metallic objects buried in the ground, walls, or most any other place where your eyes can't see. This detector does the seeing for you.

Adjust the meter for mid-scale reading with control R2 (see Fig. 5-7). Swing the search head as close to the ground as possible as you walk very slowly forward. When the meter pointer deflects either up or down, metallic objects should be directly beneath the center of the search head. The depth in which this detector will indicate a "find" depends upon the size of the object, but you should have no trouble at all locating dime-size objects an inch or two under the ground.

The construction of this detector is essentially the same as ones previously discussed; that is, build the search head out of a plastic Frisbee or several paper plages; use aluminum tubing or a wooden broom handle for the handle; build the circuit board as indicated in Fig. 5-7; and connect the detector circuit to the search-head coil. Besides the mechanical parts, you will need the electronic components listed with Fig. 5-7.

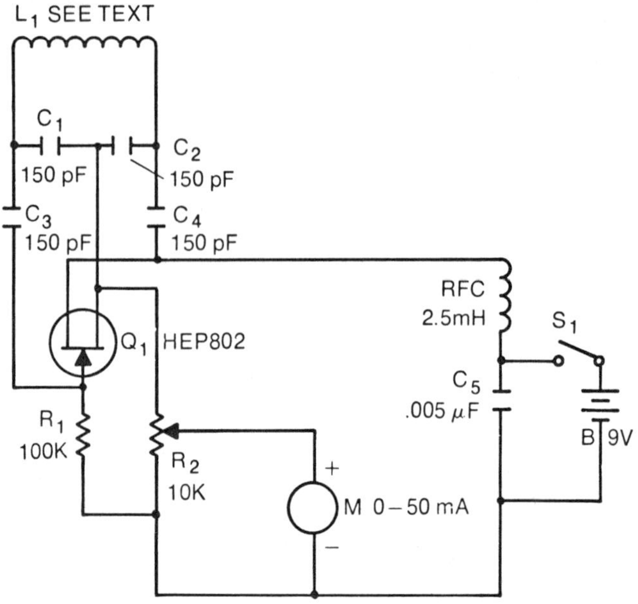

PARTS LIST

B—9 volt transistor radio battery
C1, C2, C3, C4—150 pF mica capacitor
C5—.005µF disc ceramic capacitor
L1—22 turns No. 22 AWG enameled magnet wire scramble wound on a 6-inch diameter form
M—0 – 50 mA meter
Q1—HEP802 transistor (Radio Shack 276-2035 or equal)
RFC—2.5 mH RF choke
R1—100K 1/2 watt carbon resistor
R2—10K linear potentiometer
S1—single-pole, single-throw switch

Fig. 5-7. Beat frequency oscillator type detector utilizing a 0 – 50 mA meter to detect metal objects.

Metal-mineral detectors have been manufactured in the United States since before the beginning of World War II, but during this time, only two basic types of detectors have been developed. One is the beat frequency oscillator type and the other is the transmitter-receiver type. A beat frequency oscillator detector is a constant volume audio, frequency change, all purpose, metal-mineral detector while a transmitter-receiver detector is a fixed audio frequency, loudness-change metal detector.

Of the two types, a carefully designed and constructed beat frequency oscillator type is the only detector capable of performing all phases of treasure hunting, and the circuits described in this chapter are of this type. There are, however, kits available from the major electronic supply houses which will enable you to build your own transmitter-receiver type (sometimes refered to as an "induction balance" system).

While the beat frequency oscillator type gives a change in pitch from a constant tone, the transmitter-receiver types normally do not sound until a piece of metal enters the field. Then the tone grows louder as the search head gets closer to the object. Most of these kits sell for around $70 to $90 and include a built-in speaker, headphones, and a panel meter.

Chapter 6

Building Metal Detectors From Kits

Many consumer electronics and kit manufacturers—Radio Shack, Heath Co., etc.—offer both BFO (Beat Frequency Oscillator) and TR (Transmitter-Receiver) metal locator kits. While a little bit more time-consuming to construct than the circuits given in this book, many of these kits offer high-quality metal locators—especially the TR models—which can be built by the hobbyist at a fraction of the off-the-shelf cost.

HEATHKIT

The Heathkit Model GD-348 Deluxe Metal Locator, for example, is a highly sensitive instrument for finding buried or hidden metals. Because it is simple to operate, anyone—amateur, hobbyist, or professional—can use it to pinpoint metal objects with accuracy. This lightweight unit is designed for comfortable handling to provide many hours of exciting treasure hunting, even by youngsters and can be built in an evening or two by almost anyone.

Separate SENSITIVITY and NULL controls are provided for unlimited versatility in varying geological locations. A speaker and a meter are provided to give you both a visual and audible indication when metal is detected. An output jack for headphones is also provided for use in areas where the surrounding noise level is high. In addition, the search coil

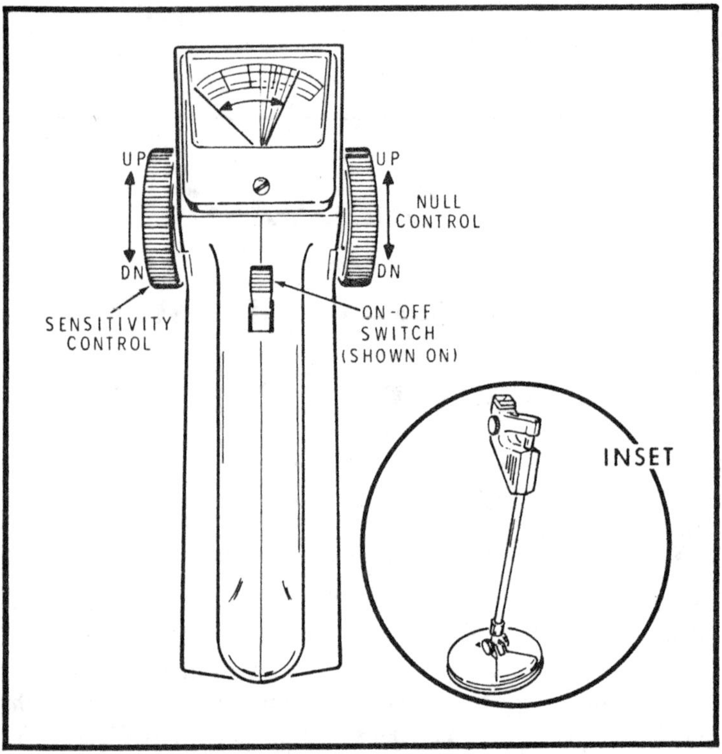

Fig. 6-1. Drawing showing the location of the controls on the Heathkit TR metal detector.

housing is designed for submersible operation in up to two feet of water (more if you detach the search head and install a longer cable).

Rugged construction and versatility of operation make this deluxe metal locator dependable and a pleasure to use. New uses will be found each day to add to your enjoyment and adventure.

Testing

Once the unit is partially assembled, according to the instructions provided with the kit, you will want to make the initial tests as follows, referring to Fig. 6-1 for the following steps:

1. Position the Deluxe Metal Locator so the search coil is positioned as shown in the inset drawing, and away from any metal or metallic objects.

2. Rotate the NULL control to the center of its rotation.
3. Slide the ON-OFF switch to the ON position. You may or may not hear a tone from the speaker. NOTE: The left handle-half is lettered with MAX, which indicates the control direction for maximum sensitivity. UP and DOWN will be used as a convenient reference in some of the following steps instead of the MAX indication.
4. If you do *not* hear a tone, rotate the SENSITIVITY control *down* until a meter indication of 6 (six) is obtained. If you hear a tone, rotate the SENSITIVITY control either up or down until you obtain a meter indication of 6.
5. Slowly rotate the NULL control in either direction until you obtain a minimum meter indication (null). NOTE: It is normal if the meter needle drops and remains at zero during some of the control rotation.
6. Alternately rotate the SENSITIVITY control *down* for a meter indication of 6, and then rotate the NULL control for a minimum meter indication. When rotating the NULL control only *increases* the meter indication, the best null has been reached.
7. Rotate the SENSITIVITY control so the pointer indicates halfway between 0 and 2 on the meter.
8. Slide the ON-OFF switch to OFF.

Operational Check

Refer to Figure 6-2 for the following steps:

1. Place a dime on a nonmetallic surface where the search coil can be passed over it.
2. Slide the ON-OFF switch to ON.
3. Hold the search coil approximately 2 inches above the nonmetallic surface and gradually pass it over the dime. Refer to Figure 6-2 for meter indications and corresponding dime positions.
4. Slide the ON-OFF switch to OFF.

Operation

This locator is designed for easy operation, but it may require a small amount of practice before you can get the best results and locate buried small metal objects accurately. Read

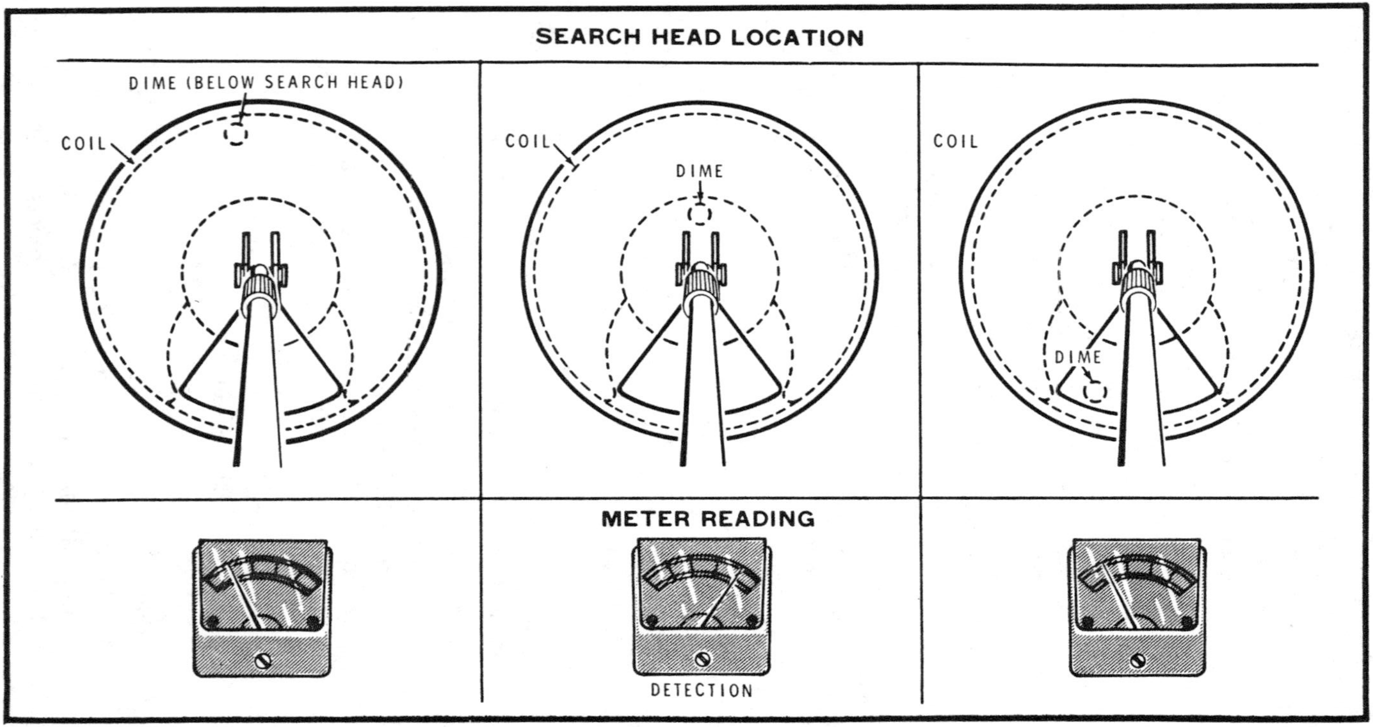

Fig. 6-2. Search head location and corresponding meter reading of a Heathkit TR metal locator.

the operation section of its accompanying manual carefully to obtain a greater understanding of the locator.

Locating small metal in tall grass, wet ground, and especially in wet gravel or sand, may prove difficult. For best results, the ground should be fairly dry.

The search coil may be rotated 90° and folded flat against the shaft. Forcing the search coil to rotate if the shaft is not fully *collapsed* will damage the shaft liner. When the shaft is fully collapsed, the metal locator can be carried in a carrying case that comes with the kit.

In order to prepare the metal locator for use:

1. Loosen the shaft locknut and extend the telescoping shaft to a comfortable length when the search coil is about 2 inches above the ground. Retighten the shaft locknut.
2. Position your Deluxe Metal Locator so the search coil is folded down and away from any metal or metallic objects.
3. Rotate the NULL control to the center of its rotation.
4. Slide the ON-OFF switch to the ON position. You may or may not hear a tone from the speaker.
5. If you do not hear a tone, rotate the SENSITIVITY control down until a meter indication of 6 (six) is obtained. If you hear a tone, rotate the SENSITIVITY control either up or down until you obtain a meter indication of 6.
6. Slowly rotate the NULL control in either direction until you obtain a minimum meter indication (null). NOTE: It is normal if the meter needle drops and remains at zero during some of the control rotation.
7. Alternately rotate the SENSITIVITY control down for a meter indication of 6, and then rotate the NULL control for a minimum meter indication. The best null has been reached, when rotating the NULL control either way only increases the meter indication.
8 Rotate the SENSITIVITY control so the pointer indicates halfway between 0 and 2 on the meter.
9. Slide the ON-OFF switch to OFF.

How To Use The Metal Locator

1. Place a dime on a nonmetallic surface where the search coil can be passed over it. NOTE: The most

sensitive spot on the search coil is just forward of the shaft thumbnuts. When this spot on the search coil is directly over the buried metal, the strongest indication will be obtained, thus pinpointing the location of the metal.

2. Slide the ON-OFF switch to ON.
3. Readjust the SENSITIVITY and NULL controls while holding the search coil about 2 inches above the surface to be searched.
4. Hold the search coil approximately 2 inches above the nonmetallic surface and gradually pass it over the dime. Figure 6-2 illustrates the meter indication as a small coin is detected. The audio tone will also increase. Headphones may be used to provide a more sensitive means of detecting this increase. The speaker is automatically disconnected when headphones are used.
5. Slide the ON-OFF switch to OFF.
6. When you search for small metal objects, such as a coin, or for deeply buried metal, it is best to use the signal meter—which provides a more sensitive indication than the audio signal.
7. Practice finding various sizes and types of metal you have planted. The better you become acquainted with your Deluxe Metal Locator, the easier it will be for you to find buried metal.
8. Remove the battery when the Deluxe Metal Locator is to be stored for a long period of time.

Operating Hints

The sensitivity of a metal locator is to a surface parallel to the search coil rather than an edge. For example, a coin on edge or a pipe on end is harder to find than if it was lying flat. Metal locators are actually finding the closest surface areas rather than the bulk or mass of an object.

The search coil is sensitive to increases in the electrical conductivity of an object. An example would be that metal objects have more conductivity than the surrounding earth. As a result, a damp beach (especially a salt water beach) has more conductivity than the dry sand away from the water, or dirt in a field. These high conductivity conditions will somewhat decrease the sensitivity. This may cause you to

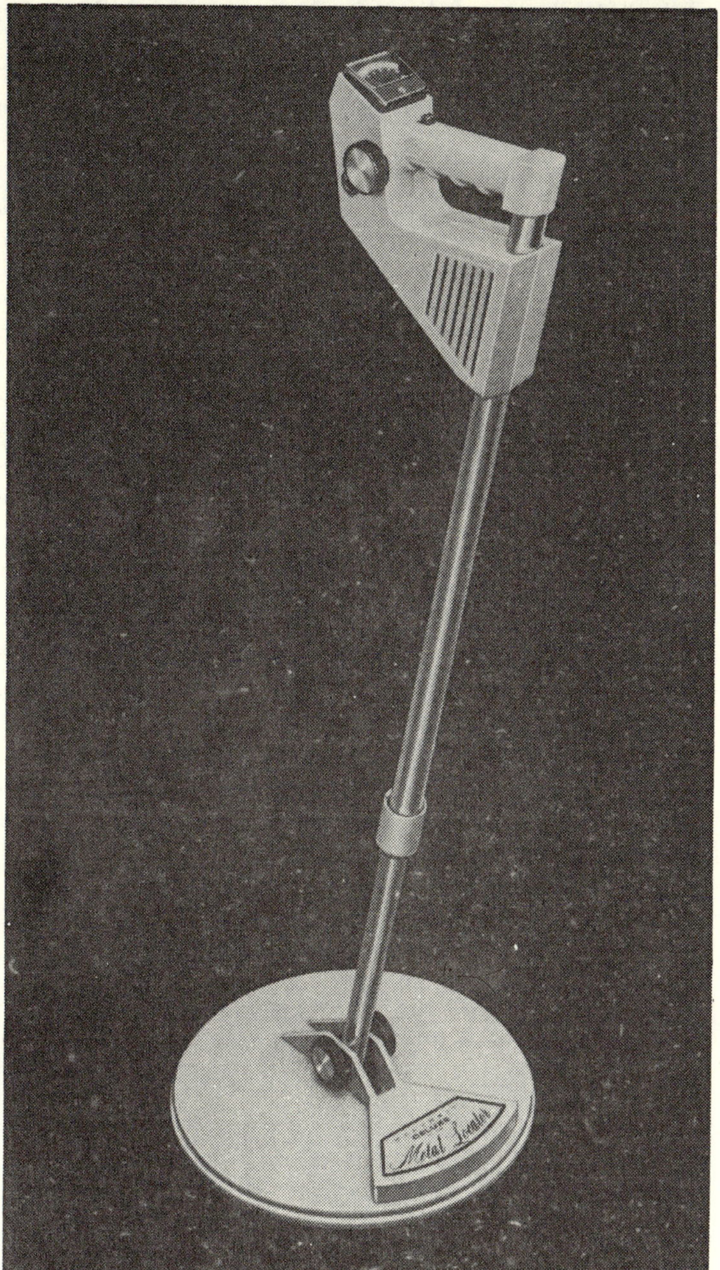

Fig. 6-3. Completed Heatkit Deluxe Metal Locator.

detect footprints on a beach or dense grass areas that retain dew.

An illustration of the completed kit is shown in Figure 6-3.

The previously described circuits in this book have been for constructing the BFO type of detectors. This type should be the choice of those who plan to engage in a wide variety of treasure hunting activities—especially in search of minerals or for use over rough terrain. On smaller items, however, inexperienced operators fail to recognize the weaker signals and therefore cannot achieve the depth as they could with the TR instrument as described in this chapter.

For those who plan to engage primarily in coin hunting with perhaps occasional hunts for relics and buried caches, the TR detector—like the one described in this chapter—would be the type to choose. It is especially easy for the beginner to learn to use in the shortest possible time. Add to this the quickness of response, its near silent tone (except when a "find" is indicated), and you have a metal detector that is hard to beat for the coin hunter.

One of the TR's disadvantages is that when it is operated over rough terrain containing ground mineralization, the instrument is bombarded with false signals which mask the true signals of buried metallic objects. The user can correct for this somewhat by tuning the detector down (desensitizing it) so that it does not react so strongly to the mineralization. Unfortunately, this also means passing up many finds which may be highly valuable.

There are several other electronic treasure hunting items that can be obtained in kit form and assembled at home in your spare time. Such items will include electric magnets for "fishing" on the bottom of lakes for valuables, or depth sounders to be able to "see" the bottom of a lake or cover, etc. Besides the fun derived from assemblying these kits, you will also save 40-60 percent from the off-the-shelf price.

Chapter 7

Repairs and Maintenance of Equipment

With the advent of transistors a few years ago, battery life in all electronic devices and instruments has been lengthened tremendously over the older vacuum-tube types—this includes metal detectors. Many transistor-powered detectors use batteries which last up to 100 hours or more, but still batteries remain the chief cause of metal detector problems.

To illustrate this fact, let's take a few of the more common problems that occur while treasure hunters are operating their instruments.

(1) NO SPEAKER VOLUME OR SOUND

This problem is generally caused by either bad batteries or else good batteries that are not seated correctly in their sockets. Since most of the better metal detectors have a built-in battery checker, the first step would be to check the batteries according to the instructions furnished with your instrument. The method of testing the batteries will run something like this:

Turn on the instrument, adjust the meter for proper detecting, and allow it to stay on for 5 to 10 minutes. This will ensure that you get an accurate battery test, because many bad batteries will read "good" when they are first energized, but will rapidly drop in voltage after they have warmed up.

After the batteries have warmed up for the time previously specified, activate the battery test switch, and if the battery is good, the meter needle will indicate this on the instrument scale. Sometimes extremely fresh batteries will cause the needle to move above the meter's scale, but this is normal. If the battery reads below normal, replace it with a new battery. Repeat the operation on all batteries in the pack or section.

Remember also that, just because a battery is new doesn't mean it can't be bad. The volume will read "good", but the capacity of the batteries to deliver current at the required low impedance or battery resistance is poor. Partially discharged batteries can also read "good" when they are first turned on, but may quickly discharge.

If you still have no speaker volume or sound after checking and replacing all defective batteries, perhaps your instrument is tuned wrong. Carefully rotate the tuning control knob from one "end" of the band to the other until the detector is tuned. If you are sure that the instrument is properly tuned, check to see if the search coil lead is plugged in and making good contact.

(2) POOR SENSITIVITY OR DETECTION CAPABILTIY

Again, low battery voltage or high impedance batteries could be causing the problem. Check the batteries as described previously and replace any that are defective.

The instrument might not be tuned correctly. If in doubt, re-read the manufacturer's instructions carefully to make certain you are operating the instrument correctly.

The object on which you are testing the instrument may not have been buried long enought to be detected as deep as you perhaps think it should be. Remember that an object buried for over a year will be detected deeper than a fresh one. Try testing your instrument in actual field conditions and let the objects you normally dig be your indication of the instrument's sensitivity. Extremely dry soil and mineral interference will also cause a loss in depth penetration.

(3) EXCESSIVE FREQUENCY SHIFTING OR INSTABILITY

The most common fault or cause of this problem is low battery voltage or high battery impedance. Check and replace any defective batteries as described before.

You may be trying to use your instrument before giving it sufficient time to warm up. Always allow 10 to 15 minutes of warm-up time prior to using your detector as this will allow your detector to reach its normal operating frequency at the prevailing temperature level. Batteries will also come to the "steady" voltage level during this warm-up time.

Should this not solve the problem, check for loose batteries or search coil connections; they should be tight.

Sometimes water leaks into the search coil and will cause the problem of frequency shifting. If you have been using your instrument on a hot summer's day and suddenly plunge it into cold water, the cooling effect can cause a partial vacuum inside the coil which has been known to pull water in through tiny holes in the coil cable where the cable enters the coil housing. These holes become punched in the cable when the instrument is operated in briars, cactus, or sharp rocks and the like. If this occurs, dry the coil out at a temperature of 100°F, then test the coil. Once you are certain that all the moisture has been dried out, seal the area around where the cable enters the coil housing with epoxy glue.

Operating your instrument over highly-mineralized ground can also be the cause of frequency shifting—so test your instrument in several locations before shipping it back to the factory.

(4) ERRATIC OPERATION

This is another common problem that is caused by bad or weak batteries. Batteries always build up to higher voltage levels when they are shut off and when this "high" voltage is in the operating range of the detector, the instrument will work perfectly for a few minutes after it is first turned on. However, this voltage drops rapidly, and if it drops below the required operating voltage, the detector can become erratic. This short period of normal detector operation, followed by sudden poor operation, is almost a certain sign of weak batteries.

Erratic operation could also be caused by loose battery terminals. Check and clean the battery terminals periodically and tighten them by slightly squeezing in the female part of the clips on both the battery holder and the battery—if these are the type of batteries you're using. Check the coil leads for a tight connection and also make certain that the leads are wrapped tightly around the detector stem. Loosely wrapped coil cables have been known to cause this problem in some types of metal detectors.

(5) ERRATIC-SOUNDING SPEAKER

Small pieces of dirt or other foreign material sometimes drop through the speaker cover holes and in turn cause the speaker to give off erratic sound. To remove the particles, turn the instrument upside down and shake it. If the pieces do not fall out repeat the procedure while turning the instrument through its slower beats (if of the BFO type). The vibration of the speaker sometimes causes the objects to fall out. If the particles are iron, the speaker magnet may attract them sufficiently so that they do not fall out. In this case, hold a low power magnet up near the speaker when the instrument is held upside down with the instrument beating at a slow rate.

Note that in four out of the five common problems just covered, it's recommended that the batteries should be checked first. From this we can see that at least one extra complete set of batteries should be carried by the treasure hunter when taking his instrument afield. One bad battery could ruin an otherwise enjoyable weekend outing searching for treasure in some backwoods.

You will also want to carry a tube of epoxy cement with you to secure any parts that may come loose. One of the most common items to break is the plastic flange that connects the aluminum search stem to the search head. The search head can be accidentally hit against a rock or other hard substance and the impact has been known to cause the plastic shaft to break, or crack. A tube of epoxy cement and possibly some fine wire can help get you through the day until the instrument can be permanently repaired at home. One treasure hunter carved a new shaft out of wood, and secured it with epoxy and wire.

If none of the previously mentioned suggestions corrects your circuitry problems, it is suggested that you ship the instrument back to the factory for repairs—they will be better equipped to handle the job and upon the instrument's return, it will be like new.

In metal detectors that you have built yourself—either from a kit or from a design of your own—there are a few suggestions for troubleshooting these in case of difficulty. Most problems occur in kits and homemade detectors mainly because of poor connections and soldering. Therefore, many troubles can be eliminated by a careful inspection of connections to make sure they are soldered properly. Resolder any doubtful connections and be sure all the wires are soldered at places where several wires are connected.

Check each circuit board to be sure there are no solder bridges between adjacent connections—removing any found by holding a clean soldering-iron tip between the two points that are bridged until the excess solder flows down the tip of the soldering iron or gun.

Be sure each transistor is in the proper location and that each transistor lead is positioned properly and has a good solder connection to the coil.

Always check capacitor values carefully and make certain that the proper one is wired into the circuit at its proper location. Also check the polarity of electrolytic capacitors to be sure they are installed correctly.

Continue to check each resistor carefully as it is very easy to get these mixed up—especially if you're an amateur starting on your first project. It would be easy, for example, to install a 1000 ohm (brown-black-red) resistor where a 100,000 ohm (brown-black-yellow) resistor is called for. A transistor that is discolored, cracked, or shows any sign of bulging would indicate that it is faulty and should be replaced.

If diodes are used in your detector circuitry, be sure the correct one is installed at its proper location. Also be certain that the banded end (cathode) is positioned correctly.

Recheck the wiring by tracing each lead according to the schematic diagram used to build the instrument. If you have a neighbor or friend familiar with basic electronics, it may be a good idea for him to go over the circuit diagram with you. Oftentimes, someone who is not familiar with the project may notice something you have consistently overlooked.

Check all component leads connected to the circuit boards and make sure the leads do not extend through the circuit board and make contact with other connections or parts. You might want to test the circuits with an ohmmeter and a voltmeter once you are reasonably certain that all connections are properly made.

The finest metal detectors on the market today are those which boast not only advanced circuitry but advanced mechanical stability as well—both are equally important.

Therefore, you must make certain that all components and connections are firmly secured, because poor mechanical stability will result in drift of the frequency. In general, rigid mounting of all components can be obtained by using short lead lengths on the parts and by using epoxy cement to hold fast any component that tends to shake or move in any way when the unit is moved. After your circuit board is completed, give it a good firm shake while observing the circuit components. If anything moves, dab several drops of epoxy on the guilty component, and let it set until dry and hard.

Some of the inexpensive factory kits could also stand some modification prior to using them in the field. Check these to make sure that the search coil lead is firmly wrapped around the search stem or handle, as well as tight connections on both ends—at the search coil and at the circuit-board housing. You may even want to check all of the connections on the circuit board to insure a good connection and to make certain that no bridging occurs.

A summary of metal detector troubleshooting is shown in the chart in Fig. 7-1. While this chart was designed for the Heathkit TR detector, it can be used equally well for other detectors (except that the part numbers may vary).

USING TEST EQUIPMENT FOR TROUBLESHOOTING

If you have access to test equipment, the use of it may possibly save you some time and help you to locate any difficulties quicker. For example, a voltmeter may be used to check the voltages in your unit against the voltages shown in a schematic wiring diagram from which your unit was built. If so, be sure to observe the following precautions:

Be cautious when you test the transistor circuits, as transistors are much more vulnerable to damage from excessive voltage or current than many of the other electronic devices—although transistors have almost unlimited life when they are used properly.

Be sure you do not short any terminals to ground when making voltage measurements. If the probe should slip, for example, and short out a bias or supply point, it is very likely to cause damage to one or more of the transistors or diodes.

In testing transistors, a transistor checker provides the most accurate method. However, an ohmmeter may be used to determine the general condition of a transistor—provided the

DIFFICULTY	POSSIBLE CAUSE
No sound in the speaker and no meter deflection when the unit is turned on and the Sensitivity control is advanced fully clockwise.	1. Battery dead. 2. Phone jack wired incorrectly. 3. Transistors or other components incorrectly installed. 4. Search coil incorrectly connected. 5. Pickup coil leads shorted. 6. Broken lead to battery. 7. Search coil leads shorting together.
Sound is heard from the speaker and meter indicates full scale regardless of Sensitivity setting.	1. NULL control R301 not adjusted for null. 2. Search coil or pickup coil improperly connected. 3. Search coil leads shorting together.
Unable to obtain a minimum reading during adjustments.	1. NULL control R301 incorrectly wired. 2. Search coil or pickup coil improperly connected. 3. Search coil leads shorting together.
Meter does not indicate or indicates backwards. Audio circuits operate properly.	1. Phone jack wired incorrectly. 2. Meter leads reversed. 3. Q207. 4. Meter 301. 5. Search coil leads shorting together.
Poor sensitivity.	1. Weak battery. 2. Improper adjustment (refer to the Operation section). 3. Transistors Q103, Q104, Q201, Q202, Q203, Q205, and/or Q206. 4. Search coil or pickup coil have shorted windings or leads. 5. Earth is so conductive that metal objects are not detected.
Erratic response when searching at a constant height above a small area of ground containing unknown objects.	1. NULL control adjusted so ferrous metals produce a decrease in the meter indication and tone, and nonferrous metals produce an increase. Refer to the "Operation" section of the Manual.

Fig. 7-1. Troubleshooting chart

ohmmeter used has at least 1 volt DC at its probes in order to exceed the threshold of the diode junctions inside the transistors. First, remove the transistor from the circuit and then set your ohmmeter on the R × 1000 range. Before performing any tests, however, make certain that you have properly identified the emitter, base, and the collector. Continue by connecting one of the ohmmeter test leads to the base lead of the transistor and then touch the other ohmmeter lead to the emitter and then to the collector. Both readings should be the same, but may be either high or low. If one reading is high and the other low, the transistor is damaged and should be replaced. Repeat this last step with the ohmmeter test leads reversed. If all the readings are either low or high—no matter which ohmmeter lead is connected to the base—the transistor should be replaced.

Chapter 8

How and Where to Use Your Detector

Now that you have built a metal detector for your own use, you are ready to enter the fascinating world of treasure hunting. The information provided herein is offered to assist you to use your detector with ease and success, so we strongly recommend that you carefully read the following instructions and operating procedures.

TUNING PROCEDURES

The importance of properly tuning your detector cannot be over-stressed. For detector circuits working in conjunction with a portable transistor radio, first tune the radio to a weak station, then adjust the variable capacitor in the detector circuit so the oscillator "beats" against the received signal. When the search coil passes over metallic objects, the inductance of the coil is changed, thereby changing the detector's oscillator frequency and changing the note of "beat whistle" heard in the radio.

To help you better understand this, place the detector in a stable position with the search-coil bottom easily accessible and away from metal objects; then adjust the "beat" to obtain a tone. Move a coin or similar object up and down above the coil, as shown in Fig. 8-1, and listen for a variation in sound. In all probability, the tone will at first sound lower for a brief second and then rapidly climb higher in pitch—depending upon

Fig. 8-1. Procedure for making air test using a coin or lead bullet.

the size of the metal object; usually the larger the object, the higher the pitch. This is known as the air sensitivity test. In the simpler circuits, your hand alone will probably cause the pitch to change, so with them, it may be best to place the coin on a flat nonmetallic surface, and bring the search coil over the object. Regardless of the sound your particular metal detector happens to give, familiarize yourself with this sound and how it changes when near a metal object.

Your next step would be to lower the search coil to the ground (away from any metal), keeping the search coil as close as possible to the ground without touching it. It may be necessary to readjust the tuning slightly since the initial tuning spot may have been different in elevation or mineralization and a large tone change may occur, either higher or lower. In either case, you will want to retune to a comfortable listening range. You are now ready to hunt.

Place several different size coins on the ground (away from any other metal objects), and swing the search head over each one about an inch above them. Listen carefully for the sound that each pass makes. Then raise the search head approximately 2 inches above the coin; again listen to the change in sound as the search head passes over the coins. Continue raising the detector—an inch at a time—until the whistle becomes too faint to hear any change. A slight adjustment of the trimmer may increase the sensitivity of the detector, allowing you to detect the coin deeper in the ground or farther away from the search head. The maximum distance that you can detect a coin while performing this "air test" will be decreased approximately 30% if the coin is buried in the ground. In fact, it wouldn't hurt anything to bury a few coins at different depths to become accustomed to the sound of the beat whistle at these depths.

FIELD OPERATION

It is important to select a rod stem that you find comfortable while searching with the detector in order to lessen body fatigue during the search. The operator should grasp the rod in such a manner that will allow him or her to stand in a perfectly normal erect manner, with the operating arm in a normal extended position at the body's side. There should be no need to hold the arm at an angle to support the instrument up or hold the body in a stooped position. Movement of the detector from left to right should be accomplished with arm movement and the swinging movement of the body while you walk slowly forward.

When scanning an area, sweeps should be made in an arc from side to side in front of the operator, keeping the search coil as close to the ground as possible. Each sweep should overlap the previous one by at least 3 inches to ensure that potential "finds" are not missed. The rate at which you can swing the detector from side to side will vary with the type of detector you happen to own. However, an average sweep speed should be somewhere around 1 foot per second for beginners. Speed can be increased as your skill in identifying signal responses increases. Try to be systematic when covering an area; that is, walk in a straight line to insure no spots are missed, then walk back along the side of your first path. Continue this procedure until the entire area is covered.

Again, you may want to bury various sizes of test targets to further familiarize yourself with the beat sound of various items. Try burying coins, bottle caps, a tin can top, and similar items 1 inch deep in a smooth area and remember their position. Try to distinguish between the different size items by listening to the difference in the beat-tone sound.

Now that these various tests have been competed, it is highly recommended that you practice over and over again for familiarization of all the characteristics of your particular detector. This will instill confidence in the operator before going out into the field on a treasure hunt.

After contacting a target on a side to side sweep motion, come back across the target several times and stop when the loudest signal is heard—in some types of detectors—or when the highest pitch of the beat tone is heard. Your "find" should then be directly in the center of the coil on a left to right plane. Now move the loop in a forward and backward direction until the signal is a sound which you have come to know as a "find". You should now have the target approximately centered. However, targets that are tilted on an angle under the ground or on edge will probably be off center either left or right, but usually no more than an inch or two.

DIGGING YOUR FIND

For small objects, a blunted ice pick, long shanked screwdriver, scratch awl or any other long bladed tool that can be pushed into the ground easily will make a good "find" probe tool. Probing for your find is very important because if you do not dig your find properly, you can mar the object or cause an unsightly hole in the ground. This latter result certainly would not promote good relations with the landowners for future treasure hunting. In fact, in many areas of the United States, treasure hunting has become illegal in national parks, city parks, and similar areas due to the abuse of the grounds by treasure hunters.

Once you have centered your target, mark the spot with your eye, and probe straight down gently with your selected tool until it comes in contact with the object. Once you have made contact, remove the probing tool, and insert back under the object with your probing blade at a 45° angle; then very carefully, pry upward, break the sod, and loosen the object. A little more loosening and your probe blade should pop the object to the surface. Replace the dirt and sod in its original position and then step on it lightly. The ultimate objective in digging for objects is to leave the sod as it was before you started digging.

In non-public areas or out in the hills, a little garden trowel or spade can be used to recover the larger objects. However, the same rule of filling holes and removing trash applies.

CHOOSING AN AREA

Before discussing the probable areas where treasure may be found, there are a few rules and regulations that should govern the use of metal detectors regardless of where you intend to search.

When searching on private property, always obtain permission from the property owner before you begin your search. If you are tracing down or expect to locate a large or valuable "find" on the property, it is wise to make the individual property owner aware of your intentions before beginning the search. Sometimes, where areas would normally be restricted to you, through proper identification of you and your equipment—together with some diplomacy—you may be able to gain permission to hunt for "treasure" where others have failed. You may also find that the property owner may have some super ideas that could assist you in locating more valuable "finds" than you otherwise could not without his assistance. Then, you and the property owner should be able to work out some type of an arrangement whereas you could split the "find" on a 50-50 basis—or something of similar convenience.

Once the permission is granted, always have respect for the property itself. This was stated previously, but it's worth mentioning once again. Never leave any type of hole or surface mar on the property, as you may wish to search there again. Also keep in mind that, if you do leave unsightly remains, it is quite possible that the next person behind you will not be granted permission to hunt for treasure on the property. Then too, you may even be the next person someone else has left a mess for!

Currently, it is prohibited to use a metal detector in national parks, within the boundaries of national monuments, state parks and other designated historical sites, so keep clear of these areas. Certain violators have not only been fined, but have had their detectors, automobiles, and other belongings confiscated as well! If you are not sure of existing city ordinances in your area, you should contact your local officials concerning the regulations on treasure hunting and digging in the various areas.

The best areas for locating single coins, rings, jewelry, and the like, are places where there have been large gatherings of people out of doors on a regular basis over a period of years—the older the site, the better the chance of finding valuables. Examples are old carnival sites, recreational parks, city parks, swimming holes, playgrounds, race tracks, school grounds, parking strips, and many more. These areas will offer many targets for you to get acquainted with your metal detector, and you will certainly find one or more of them within a short distance from your home.

An old field on the outskirts of your town may now be a pasture for a herd of cattle, but years ago it could have once been a regular meeting place for tent shows or carnivals. Check with the older citizens of your community and find out where these gatherings took place.

Obtain books from a nearby library about the history of your area. The ones who have the greatest success at finding valuable treasures are those who take the time to do some research *before* they begin hunting.

Old abandoned home sites are excellent places to find valuables of all types. If you can't find the location of such a place from old country maps, again ask the older citizens around the area. There are many old abandoned home sites all over the country in rural areas, and it is a hard matter indeed to search around one of these sites and come away empty handed.

In the old days, many of the country people didn't believe in banks, and most had some safe place—right on their own property—in which to hide their money and other valuables—behind a loose stone or brick in the chimney—under a floor board—buried at the corner of a particular outbuilding. There is a good chance that much of this treasure still remains at the sites of these old homes. The owners could have died, telling no one the whereabouts of their lifetime savings. A person with a reasonably good metal detector and a little patience could uncover a lot of these buried treasures if the location of the home sites could be located.

Happy treasure hunting!

Appendices

Appendix A
List of Equipment Dealers

The following dealers specialize in metal detectors and other treasure hunting equipment. If you don't find exactly what you need at the dealer closest to your home, they will most certainly be able to direct you to the proper place.

ALABAMA

G & B Detector Sales
BMK Building
1378 Hwy 31 So.
Alabaster 35000
(205-663-4573)

Birmingham Coin & Relic
Detector Sales
113 24th Ave. N.W.
Birmingham 35125
(205-854-0341 or 854-3364)

Dan Owens
1317 S. 11th St.
Gadsden 35901
(205-546-6561)

ALASKA

Stewart's Photo Shop
531 Fourth Ave.
Anchorage 99501
(272-8581)

ARIZONA

McCowan Enterprises
824 E. Broadway Rd.
Phoenix 85040
(602-276-6774)

ARKANSAS

W. W. Mosely
P.O. Box 7
Camden 71701
(501-836-5314)

Herring Electric Co.
1217 W. Hillsboro St.
Eldorado 71730
(501-862-3928)

Bill's Detectors
5623 "R" St.
Little Rock 72207
(501-666-6355)

L. L. Lincoln
Rt. 1
Rogers 71756
(501-636-6867)

CALIFORNIA

General Electronic Detection
16238 Lakewood Blvd.
Bellflower 90706
(213-633-t330)

Aurora Prospector's Supply
6286 Beach Blvd
Buena Park 90620

Rencher Welding & Machine Works
560 S Third St.
Chowchilla 93610
(209-665-4219)

Escondido Coin Shop
111 North Broadway
Escondido 92025
(714-745-1613)

Gene Rolls
Hwy. 32 at Sugar Pine
Forest Ranch 95942
(916-342-4829)

Fresno Hobby & Crafts
3026 N. Cedar
Fresno 93703
(209-226-4880)

Central Cal Electronics
600 East St.
Hollister 95023
(408-637-1704)

The We Shop
221 Main St
Huntington Beach 92648
(714-536-4700)

South Bay Coins
639 Ninth St.
Imperial Beach 92032
(714-423-2551)

Fumble Fingers
3384 Mt. Diablo Blvd.
Lafayette 94549
(415-284-7406)

White's Detectors
Arthur Apodaca
3141 E. Highway 140
Merced 95340
(209-723-0636)

Gold Nugget Miner's Supply
1302 9th St.
Modesto 95354

Gems Galore
240 Castro
Mountain View 94040
(415-968-8707)

The Coin Shop
1516 Third St.
Napa 94558
(707-255-8166)

Treasures Unlimited
526 66th Ave.
Oakland 94621
(415-638-6352)

N & A Belle Eelctronics
126 West Holt Blvd.
Ontario 91761
(714-984-9119)

Buddy Sales
18552 Sherman Way
Resenda 91335
(213-342-5113)

Pioneer Recoveries
35210 Audubon Place
Riverside 92501
(714-682-4302)

Mother Lode Diving Shop
2001 Capitol Ave.
Sacramento 95814
(916-443-3237)

Coin Chest Coin Shop
545 West Baseline
San Bernardino 92410
(714-885-5711)

Gems & Things Assoc.
2493 San Diego Ave.
San Diego 92110
(714-297-2672)

San Diego Coin Exchange
3784 30th St.
San Diego 92104
(714-296-3131)

Arts & Hobbies
12334 Kagel Canyon Rd.
San Fernando 91342
(213-896-9477)

Mining & Lapidary
131 - 10th St.
San Francisco 94101
(415-626-6016)

Dunall's Outdoor Supply
605-7 State St.
Santa Barbara 93010
(805-963-5319)

Hans Auto Supply
555 Charles St.
Seaside 93955
(408-394-6044)

The Treasure Cave
18 W. Bradford St.
Sonora 95370
(209-332-5856)

Wright's
2405 N. El Dorado St.
Stockton 95204
(209-466-4351)

COLORADO

Suack Enterprises
5190 N. Nevada
Colorado Springs 80907
(303-598-0701)

The Prospector's Pick
2 Ruxton Ave.
Manitou Springs 80829
(303-685-1457)

Dan & Reg's Treasure Instruments
P.O. Box 1697
Pueblo 81003
(303-566-0698)

Exanimo Establishment
Segundo 81070

Metal Detectors Sales
6117 W. 38th Ave.
Wheat Ridge 80033
(303-422-4566)

CONNECTICUT

Edward Perchaluk
304 Circle Dr.
Stratford 06497

FLORIDA

Underwater Unlimited
216 Palermo Ave.
Coral Gables 33134
(305-445-7837)

Harry's Pawn Shop
519 Main St.
Jacksonville 32202
(904-353-6971)

Edgewood Coin Shop
934 South Edgewood Ave.
Jacksonville 32205
(904-389-0013)

Palm Plaza Camera
713 N. 14th St.
Leesburg 32748
(904-787-4661)

Kellyco
1811 Chinook Trail
Maitland 32751
(305-647-2285)

Tamiami Gun Distributors
2975 S. W. 18th St.
Miami 33135
(305-642-1941)

Kenin, Inc.
1834 N.E. 163rd St.
N. Miami Beach 33162
(305-949-7681)

Wen-Co Enterprises
6874 80th Ave. N.
Pinellas Park 33565
(813-546-9264)

Dan Byrd Fish Camp
At New Pass Bridge
Sarasota 33577
(813-388-2488)

Treasure Shack
3934 Britton Plaza
Tampa 33611
(813-833-9841)

GEORGIA

S & H Metal Detector Sales
2052 Upper Ridge Rd. S.E.
Dalton 30720
(404-226-3981)

Finders Co.
225 Upland Rd.
Decatur 30030
(404-377-0974)

Ernest M. Andrews
2755 Sylvan Rd.
East Point 30344
(404-766-8141)

The Stamp Shed
1115-D Watson Blvd.
Warner Robins 31093
(912-923-6159)

J.C. Ballentine
Hatcher Point Mall
Waycross 31501

IDAHO

Mountaineering Outfitter
62 N. Main, Box 116
Driggs 83422
(208-354-2222)

Outdoor Hobby Supply
2416½ E. Main
Lewiston 83501
(208-743-1768)

Roy Norstrom
339 Chateau Thierry
Box 419
Soda Spring 83276
(208-547-4166)

ILLINOIS

Ren's Treasure Trove
1101 N. Livingston
Bloomington 61701

Wheat State Electronics
12203 Vincennes Ave., Apt. 24
Blue Island 60406
(312-371-70896)

Jerry's Gun Shop
9220 Ogden Ave.
Brookfield 60513
(312-485-5200)

Gene M. Blaker
2903 West John
Champaign 61820
(217-356-0918)

Harry's Treasure Shack
322 W. State St.
Cherry Valley 61016
(815-332-5157)

Frontier General Store
1066 W. Harrison
Decatur 62526
(217-877-9199)

Automatic Engineering
1630 Ogden Ave.
Downers Grove 60515

B-T Electronics
31. S. McLean Blvd.
Elgin 60120
(312-742-3090)

White's Metal Detectors
302 Poplar St.
Harrisburg 62946
(619-253-5131)

The Treasure Trove
727 S. Finley Ave.
Lombard 60148
(312-629-3980)

N & D Detector Sales
2110 N. Richmond Rd.
McHenry 60050

Dal-Mar Electronics
5815 W. Warren St.
Morton Gove 60053
(312-966-9687)

Rev. John J. Costas
3116 11th Ave. "A"
Moline 61265
(309-797-3098)

Pioneer Treasure Detectors
P.O. Box 55
123 Dee St.
Moro 62067
(618-377-8655)

Orland Treasure Finders
14231 S. Union Ave.
Orland Park 60462
(312-349-2078)

Dee's Beauty Shop
206 Reservoir Rd.
Pekin 61554
(309-346-4377)

B & E Sales
1009 South Tonti Circle
Peoria 61605
(309-637-1570)

Mid-West Treasure Detectors
505 South 8th St.
Quincy 62301
(217-223-4723)

Tom's Pool Center
809 N. Green Bay Rd.
Waukegan 60085

INDIANA

O-D Western Store
Rt. 5
Decatur 46733
(219-724-2097)

Krull's Hobby Shop
414 E. Washington Blvd.
Ft. Wayne 46802
(219-422-4429)

C & E Treasure Seekers
16340 Fairfield Ln.
Granger 46530
(219-272-3750)

J & J Coins
7019 Calumet AVe.
Hammond 46324
(219-932-5818)

Hansen's Sports
3750 Ridge Road
Highland 46322
(219-838-7495)

Metal Locating Equip.
2916 W. 62nd St.
Indianapolis 46268
(317-251-8472)

Pollard Coin & Stamp
Supply Co., Inc.
5220 East 23rd St.
Indianapolis 46218
(317-547-1306)

IOWA

Richard Cross
Box 90
Baxter 50028
(515-227-3391)

George & Dorothy Morgan
2003 Walnut St.
Cedar Falls 50613
(319-266-0271)

Geode Industries, Inc.
106-108 W. Main St.
New London 52645
(319-367-2286)

Spragg Electronic Service
1501 W. 13th St.
Davenport 52804
(319-323-7444)

The Treasure Chest
1525 S. Concord
Davenport 52804
(319-322-9355)

KANSAS

Carl Clare
911 3rd Ave.
Dodge City 67801
(316-225-4701)

Bob Ruth
Everest 66424
(913-548-3385)

Epp's Coin Shop
112 S. Main
Pratt 67124
(316-672-6181)

Bonta's Treasure Shack
617 Chase
Wichita 67213
(316-942-6965)

Mid-Western Research & Supply
6323 Scottsville
Wichita 67219
(316-744-0668)

KENTUCKY

Lewis Eddleman
3320A Wood Valley Ct.
Lexington 40502
(606-266-0980)

A.F. Waller
10800 Marcitis Rd.
Louisville 40272
(502-937-8998)

Cecil & Red
2529 W. Market St.
Louisville 40212
(502-778-2119)

LOUISIANA

Clark's Detector Sales
3233 Illinois Ave.
Kenner 70062
(504-443-1285)

Woody's Coin Shop
520 Miller Ave.
Westlake 70669
(318-439-9616)

MARYLAND

Bay Side Treasure Detectors
422-430 E. 32nd St.
Baltimore 21218
(301-235-4412)

South Mountain Sports
Rt. 2 Box 344A
Keadle Rd.
Boonsboro 21713
(310-739-5384)

Frank White's Detectors
408 Arbor Drive
Glen Burnie 21061
(301-768-3157)

Martin's Metal Detector Sales
601 Robinhood Rd.
Havre de Grace 21078
(310-939-1417)

Albert M. Schone
430 Leyton Rd.
Reisterstown 21136
(301-833-6124)

Robert F. Walters
Love Point Rd.
Stevensville 21666
(301-632-6397)

MASSACHUSETTS

E & D Electronic Sales
83 Parker St.
Agawam 01001
(413-786-7190)

Ronimac Firearms Co.
530 Chaplin St.
Ludlow 01056
(431-589-9096)

Larry Violette
Box 74
Rehoboth 02769
(617-252-4497)

MICHIGAN

Johnston's TV
Box 161
Baraga 49908
(906-353-6858)

Wayside Sports Center
1720 W. Brady Rd.
Chesaning 48616
(517-845-3030)

Detectors
5612 Croissant
Dearborn Heights 48122
(313-274-4186)

Grant's Book Store
601 Bridge St. N.W.
Grand Rapids 49504
(616-458-6480)

Treasure Hunter's Supply
1949 Collins S.E.
Grand Rapids 49507
(616-243-0216)

Electronic Metal Detectors
525 Riley St.
Holland 49423
(616-396-5692)

Electronic Center
1010 Washington St.
Marquette 49855
(906-228-9670)

Marshall Artifact Recoveries
205 W. Michigan Ave.
Marshall 49068
(616-781-4575)

Treasure Land
38554 Groesbeck Hwy.
Mt. Clemens 48043
(313-469-0600)

MINNESOTA

Phelps Comm. Ctr.
Rt. 1 Box 58
Albert Lea 56007
(507-373-1111)

Treasure Sport Specialty Co.
1208 Park St.
Anoka 55303
(612-421-8971)

Mid-West Detectors
11120 Vincent Ave. So.
Bloomington 55431
(612-881-5254)

Schilling Coins
104 E. Superior St.
Duluth
(218-722-8065)

Sage Christian Enterprises
321 W. First St.
Park Rapids 76470
(218-732-5907)

Minnesota Prospector's Supply
Rt. 4
Red Wing 55066
(612-923-4728)

Betlach Jewelers
8432 Highway 7
Knollwood Plaza
St. Louis Park 55426
(612-935-4308)

Herman Hafstad
Rt. 1
Willmar 56201
(612-235-5224)

MISSISSIPPI

Mid-South Treasure Barn
4226 Hwy 80 West
Jackson 39209
(601-922-1663)

Hobbies Unlimited
1219 Nelle St.
Tupelo 38801
(601-842-6031)

MISSOURI

R & R Stamp & Coin
Flat River 63601
(314-431-4746)

Bob's Treasure Shack
8108 E. 80 Terrace
Kansas City 64138
(816-358-4928)

Clevenger's Hobby
8206 North Oak
Kansas City 64118
(816-436-0697)

Ray George Co.
3456 S. Grand Ave.
St. Louis 63118
(314-776-4568)

Paul Snyder
202 Owens
Smithville 64089
(816-873-3339)

Radford Jewelers
1864 S. Glenstone
Springfield 65804
(417-881-7308)

Frank Sales & Service
711 S. Jefferson
Webb City 64870
(417-673-2650)

MONTANA

J & L Coinfinders
2102 Miles Ave.
Billings 59103
(406-656-6569)

Mid-West Welding & Machine Shop
Box 1115
2320 North 7th Ave.
Bozeman 59715
(406-487-5417)

Farmer's Electric Supply
1008 20th St. So.
Great Falls 59401
(406-453-5501)

Electronic Parts
1030 S. Ave. W.
Missoula 59801
(406-543-3119)

H & B Welding Service
808 3rd Ave.
Roundup 59072
(406-323-2532)

NEBRASKA

The Spartan Shop
335 N. Williams
Fremont 68025
(402-721-9438)

Canfield's
2415 Cuming
Omaha 68131
(402-342-1517)

L.P. Enterprises
Box 46
Sprague 68438
(402-794-5730)

NEVADA

Karl's CB Sales
1201 Stewart
Las Vegas 89101
(702-382-5011)

Nevada Coin Mart Supply
2409 Las Vegas Blvd. So.
Las Vegas 89105
(702-732-1375)

NEW HAMPSHIRE

Don Wilson Sales
83 S.State St.
Concord 03301
(603-224-5909)

George F. Streeter & Co.
265 Wash. St.
Keene 03431
(603-357-0229)

NEW JERSEY

General Sales Co.
10 Humphrey St.
Englewood 07631
(201-568-5563)

Geo-Quest
78 Kenzel Ave.
Nutley 07110
(201-667-8170)

Sherando Corp.
61 W. Kings Hwy.
Mt. Ephraim 08059
(609-931-1113)

The Treasure Cove
1055 S. Clinton Ave.
Trenton 08611
(609-394-8388)

The Treasure Chest
Mt. View Electric Co.
866 Rt. 23
Wayne 07470
(201-694-0777)

NEW MEXICO

Raydon Electronics
820 Sullivan
Farmington 87401
(505-325-4883)

NEW YORK

Tools Unlimited
2106 Union Blvd.
Bay Shore 11706
(516-581-4883)

Bigford Brothers
44 Clyde St.
Earlville 13332
(315-691-3563)

Panna's Electronic Sales
P.O. Box 167
Geneva 14456
(315-789-0809)

Bond Electric Co.
2 Leich Circle So.
Glen Cove 11542
(516-676-1380)

Trade Mart Enterprises
94 Keller Ave.
Kenmore 14217
(716-875-0951)

Wheels Afield Travel Trailer
Rt 3 Box 142
Kingston 12401
(914-331-5687)

Harbor Coins & Stemps
134 Mamaroneck Ave.
Mamaroneck 10543
(914-698-5671)

Kay's Detector Sales
Rt. 9 Malta
Mechanicville 12118
(518-399-4961)

C-T Detectors
293 W. 1st St.
Mt. Vernon 10550
(914-668-0518)

M. Friedlander Hardware, Inc.
146 W. 54th St.
New York 10019
(212-247-7774)

Detection Enterprises of Rochester
366 Brookview Drive
Rochester 14617
(716-338-2588)

Doc Dave's TV Service
54 Stockton Ave.
Walton 13856
(607-865-5188)

NORTH CAROLINA

Pete's Electric
P.O. Box 3
Dickinson 58601
(701-225-5025)

Detector Sales Co.
131 Stetson Dr.
Charlotte 28213
(704-596-1953)

Science Hobbies, Inc.
2615 Central Ave.
Charlotte 28205
(704-375-7684)

Carson's Coin Shop
122 N. Center St.
Statesville 28677
(704-873-9671)

OHIO

Bennett Detector Sales
203 E. Market St.
Akron 44308
(216-367-9664)

Cutcher's Brownhelm Store
1605 N. Ridge Rd.
Amherst 44001
(216-988-4110)

Red's Sport
Rt. 39 & 62
Berlin 44610
(316-893-2545)

Aluebelhart's Sport Shop
2819 Fifth St. W.
Canton 44708
(216-455-7837)

Kilian Detector Equip.
1031 Spring Rd.
Cleveland 44109
(216-398-4779)

Harbor Coin
215 3rd St.
Fairport Harbor 43927
(216-354-2458)

3-E Rock-N-Treasure Ctr.
638 Chestnut Ridge Rd.
Hubbard 44425
(216-534-4482)

Klinger's Rocks N Things
411 Bowman Rd.
Lima 45804
(419-227-5294)

Knapp's Metal Detector Sales
18944 Mitchell Ave.
Rocky River 44116
(216-331-1658)

Purcell's
109 S. First St.
Tipp City 45371
(513-667-2143)

Allen Coin Shops
34 N. State St.
Westerville 43081
(614-882-3937)
and
1243 W. Broad St.
Columbus 43216
(216-376-9694)

OKLAHOMA

B & J Treasure Shack
1219 WFP Blvd.
Bartlesville 74003
(918-336-2155)

Russey's Electric
620 Stadium Dr.
Hobart 73651
(405-726-3686)

Arrowhead Supply
330 S.W. 28th St.
Oklahoma City 73109
(405-634-7128)

Hobby World
2433 Plaza Prom
Shepherd Mall
Oklahoma City 73107
(405-942-4556)

Mid-Del TV & Electronics
4518 S.E. 29th
Oklahoma City 73115
(405-677-1194)

H & H Metal Detector Sales
114 S. Wewoka
Wewoka 74884
(405-357-3137)

OREGON

Carla Kay Salvage
471 Bruel St.
Coos Bay 97420
(503-888-4015)

Oregon Sporting Goods
175 Main
Hermiston 97838

Bennington Detector Sales
5059 Bryant Ave.
Klamath Falls 97601

Hal's Car Rental
903 Adams Ave.
La Grande 97850
(503-963-3884)

Pendleton Stamp & Coin
432 S. Main
Pendleton
(903-276-6870)

D & K Detector Sales
13809 S. E. Division
Portland 97236
(503-761-1521)

Young's Sporting Goods
515 E. 2nd
The Dalles 97058
(503-296-2544)

PENNSYLVANIA

Bob's Treasure Shack
335 East 11th St.
Berwick 18603
(717-759-1783)

Milloway's Motorcycle and
Parts Shop
439 E. 8th St.
Berwick 18603
(717-759-9372)

J & D Metal Locating Equip.
Penn-Way Mkt.
Rt. 1 Mundy's Corner
Conemaugh 15909
(814-749-9411)

Gettysburg Electronics
27 Chambersburg St.
Gettysburg 17325
(717-334-8634)

Joel Guldin
Rt. 1 Snyder Rd.
Green Lane 18054
(215-234-4932)

Sealand's Metal Detectors
422 Sells Lane
Greensburg 15601
(412-834-3429)

Miller's Treasures & Coin Detectors
Rt. 1 Pettis Rd.
Meadville 16335
(814-335-4936)

James Elder
Rt. 1 Box 253-A
Mifflintown 17059

Fox Metal Detectors
313 Bridge St.
New Cumberland 17070
(717-232-4629)

Cavalier Metal Detector Sales
333 Carmell Dr.
Pittsburgh 15241
(412-941-5691)

Richard Haviland
3228 Brownsville Rd.
Pittsburg 15227
(412-882-8073)

Prospect Metal Detectors
737 2nd Ave.
Prospect Park 19070
(215-534-1685)

Mid State Sales
3410 W. College Ave.
State College 16801
(814-238-5666)

RHODE ISLAND

House of Bargains
345 Warwick Ave.
Warwick 02888
(401-781-8580)

Bob's Service
RFD 1, Box 135
Benefit St.
Greene 02827
(401-397-3143)

TENNESSEE

Hickory Valley Electric Co.
6916 Lee Hwy.
Chattanooga 37421
(615-892-0525)

C/S Detector Electronics
354 Miflin Rd. Rt. 6
Jackson 38301
(901-424-6319)

The Collector's Shop
100 Oak Shopping Center
Nashville 37204
(615-383-5996)

TEXAS

William B. Doss
3816 N. 11th
Abilene 79603
(915-672-5102)

Arlington Electronic Center
915 East Park Row
Arlington 76010
(817-261-9441)

Arlington Treasure Chest
1735 S. Cooper
Arlington 76010
(817-261-5272)

Bill's Trailer Sales
1800 New Jersey
Baytown 77520
(713-422-4322)

M & M THers Supply
Box 83
Bowie 76230
(817-872-3439)

Treasure Hound Detector Sales
400 Mitchell
Bryan 77801
(713-823-6423)

Hide-N-Seek
288 S. 101 Hwy.
Clute 77531
(265-6562)

National Treasure Hunter's League
11602 Garland Rd.
Dallas 75218
(214-328-4327)

AM Camp
5440 Will Ruth
P.O. Box 4547
El Paso 79914
(915-751-7741)

El Paso Equip. Rental
6055 Alameda
El Paso 79905
(915-778-3351)

Joe Krajca & Son
Rt. 3
Ennis 75119
(214-875-6358)

Don Gray Enterprises
1561 W. Berry
Fort Worth 76110
(817-921-2431)

Rex Grove Auto Supply
4527 E. Belknap
Fort Worth 76117
(817-838-3066)

Jen-Mar Laundry & Cleaning
2805 Weiler
Fort Worth 76119
(817-451-9005)

Research & Recovery
2803 Old Spanish Trail
Houston 77054
(713-747-4647)

J.W. & A. Assoc. Inc.
3229 W. Pioneer
Irving 75061
(214-252-7661)

Noel & Betty Murchison
201 George
Lubbock 79416
(885-2359)

Mesquite Craft & Hobby Mart
508 So. Galloway
Mesquite 75149
(214-285-2930)

Mission Drug Store
1030 Conway
Mission 78572
(512-585-1532)

The Treasure Bug Shop
Rt. 3, Box 511
Orange 77630
(713-745-1190)

Owens Detector Sales
5814 Kepler Dr.
San Antonio 78228
(434-1605)

San Antonio Mission Metal Detector Co.
2302 W. Hermosa Dr.
San Antonio 78201
(512-732-5522)

UTAH

John Bylund Plumbing & Heating
1255 E. Center
Pleasant Grove 84062
(801-785-3540)

Bryant T. Cash
2457 W. 4975 S.
Roy 84067
(801-825-7858)

Gallenson's
2209 S. Sate St.
Salt Lake City 84111
(901-328-2016)

VERMONT

Leisure Lines
Woodstock Ave.
Rutland, 05701
(802-775-0854)

VIRGINIA

Suburban Detector Sales
3169 Spring St.
Fairfax 22030
(703-273-2542)

Chancellorsville Outpost
Rt. 1, Box 170
Fredericksburg 22401

Essential Instruments
10453 Medina Rd.
Richmond 23235
(804-272-5558)

Tabb Trailer Sales
1459 Hwy. 17
Tabb 23602
(804-595-5720)

H & S Detector Center
2114 Thoroughgood Rd.
Virginia Beach 23455
(804-464-6072)

Opequon Trader
807 Berryville
Winchester 22601
(703-667-8257)

WASHINGTON

The Treasure Chest
Rt. 1, Box 39-F
Granite Falls 98252
(206-334-5263)

The Coin Cradle
320 W. Kennewick Ave.
Kennewick 99336
(509-586-4982)

Doug's Hobby &
Prospecting Shop
6908 214th S.W.
Lynnwood 98036

Pearl Electronics
1300 First Ave.
Seattle 98101
(206-622-6200)

Bob Carstensen
17 Whispering Firs
Silverdale 98386
(502-692-6882)

Bowen's Hideout
1823 S. Mt. Vernon
Spokane 99203
(509-534-4004)

B & I Coin Shop
8012 S. Tacoma Way
Tacoma 98499
(206-584-6241)

Snake River Prospectors's Supply
821 St. John St.
Walla Walla 99362
(509-525-8304)

WEST VIRGINIA

Murdock's Hobby Shop
121 N. Fourth Ave.
Paden City 26159
(304-337-2711)

WISCONSIN

Don's Treasure Hunting Supply
P.O. Box 177
Menomonee Falls 53051
(414-251-5350)

Nyland's, Inc.
1607 Main St.
Marinette 54143
(715-735-5110)

Keith Rasmussen
1436 Michigan Blvd.
Racine 53402
(414-637-6658)

Jetzer's Locksmith Serv.
1822 N. 12th St.
Sheboygan 53081
(414-457-9231)

WYOMING

Mike's Sporting Goods
Box 1505
Casper 82601
(307-234-2294)

Appendix B

Selected Publications

The following publications will prove useful to treasure hunters in search of all kinds of finds. All may be purchased from Keene Engineering, Inc., 9330 Corbin Ave., Northridge, California 91324.

PROSPECTING, MINERALS & GEMS

The Gold Divers' Handbook .... Matt Thornton
Handbook for Prospectors .... Richard M. Pearl
Diving and Digging for Gold .... Mary Hill
Gold Finding Secrets .... Edwin P. Morgan
Let's Go Prospecting .... Edward Arthur
Prospecting and Operating Small Gold Placers ...... Boericke
How and Where to Pan Gold .... Winters
GOLD—ABC's of Panning .... E. S. LeGaye
Prospecting for Gold .... W.F. Heinz
Gold Diggers' Atlas .... Robert Neil Johnson
Prospecting Hints .... United Prospectors, Inc.
Gold Fever .... DeLorenzo
Weekend Gold Miner .... Ryan
Prospecting for Gems and Minerals .... Sinkankas
Prospecting for Lode Gold .... Gregory V. Stone
Where to Find Gold in the Mother Lode .... James Klein
Geology .... Frank H. T. Rhodes
Landforms .... Adams and Wyckoff

Mining Engineer's Handbook .... Peel
Rocks and Minerals .... Zim and Shaffer
Field Guide to Rocks and Minerals .... Pough
How to Know Minerals .... Pearl
Mineral Recognition .... Vanders and Kerr
Mineralogy for Amateurs .... Sinkankas
Dan's Manual of
Mineralogy .... Revised by Cornelius Hurlbut, Jr.
Rocks and Minerals of California ...... Brown, Allan, & Stark
Exploring and Mining for
Gems & Gold in the West .... Fred Rynerson
Gold Panning with Prospector John .... Prospector John
Where to Find Gold in Southern California .... Klein
Everything You Wanted to Know About Gold .. Russel Burkett
Blue Gold (The Turquoise Story) .... Broman
The Gold Miners' Pocket Companion .... Bob Bigando, Jr.
Gold Locations of the United States .... Jack Black
Strike It Rich "Gold" Fever .... Schissler
Gold in Canada .... Street-Foster
Minerals of the World .... Sorrell
Gold in a Campground .... Keene and Thornton
The Extraction of Free Gold .... McGowen
How to Build and Operate the Sluice Box .... T.R. Glenn
Where and How to Find
Gold, Gems, or Junk .... J. Allen Kitchens
Handbook of Jade .... Hemrich
Handbook of Crystal and Mineral Collecting .... Sanborn
Gem Minerals of Idaho .... John A. Beckwith
Exploring Minerals and Crystals .... Robert I. Gait
Arizona Rock Trails .... Bitner
Guide to Fossil Collecting .... Casanova
Eastern Gem Trails .... Oles
Midwest Gem Trails .... Third Edition
Western Gem Hunter's Atlas .... Johnson
Hunting Diamonds in California .... Mary Hill
California Gem Trails .... Henry
Desert Gem Trails .... Mary Frances Strong
Fossils .... Rhodes, Zim, & Shaffer
N.W. Gem Fields Ghost Town Atlas ...... Robert Neil Johnson
Field Guide To Mexico
Oregon Gold & Gems Maps
Dana's Textbook of Mineralogy .... Ford

Gemstones of North America .... Sinkankas
Gem Trails in California .... A.L. Abbott
Roadside Geology of U.S.
Interstate 80 .... Hamblin, Rigby, Snyder & Mattews
Gold, the 79th Element .... S.R. Wendt
Bacon and Beans From A Gold Pan .... Coffey & Hoeper
Gold Panners' Manual .... Garnet Basque
Where to Find Gold in the Desert .... James Klein

LAPIDARY

Lapidary Techniques .... Craftool Press
Tumbling Techniques .... G.L. Daniel
Gem Cutting—A Lapidary's Manual .... Johm Sinkankas
Gem Cutting, The Art of .... Dake
Handbook of Gemstone Carving .... Wertz
Cabochon Cutting .... Cox
Facet Cutter's Handbook .... Soukup
Complete Guide to Micromounts .... Speckels
The Opal Book .... Leechman
Jewelry Making by the Lost Wax Process .... Greta Pack
Handbook of Lost Wax Investment Casting .... Sopcak
Precious Stones .... Max Bauer
Specialized Gem Cutting .... Jack R. Cox
Jewelry Making and Design .... Rose and Cirino
Creative Casting .... Choate
Weekend Rock Hound (Rocks to Riches) .... Ryan

OLD MINES, GHOST TOWNS & MINING CAMPS

Ghost Towns of the West .... Lambert Florin
Alaska, British Columbia, Yukon
Arizona
California
Colorado-Utah
Montana-Idaho-Wyoming
Nevada
New Mexico-Texas
Oregon
Washington
Ghost Town Trails .... Lambert Florin
Western Ghost Town Shadows .... Lambert Florin
Ghost Town Treasures .... Lambert Florin
Ghost Town El Dorado .... Lambert Florin

Western Ghost Towns .... Lambert Florin
A Guide to Western Ghost Towns .... Lambert Florin
Colorado Ghost Towns-Past & Present ...... Robert L. Brown
Ghost Towns of the Colorado Rockies .... Robert L. Brown
Jeep Trails to Colorado Ghost Towns .... Robert Brown
Southwestern Ghost Town Atlas .... Robert Neil Johnson
California Ghost Town Trails Abbott
Ghost Towns and Mining Camps of California .. Remi Nadeau
Ghost Towns of the West
Ghost Towns of the Northwest .... Norman D. Weis
Nevada Ghost Town Trails .... Abbott
California Nevada Ghost Town Atlas .... Robert Neil Johnson
Nevada Ghost Towns & Mining Camps .... Stanley W. Pahner
Nevada Ghost Towns .... Raymond C. Browne
Old Mines of California & Nevada .... Raymond
Mines of Julian .... Helen Ellsberg
Mines of the San Gabriels .... John W. Robinson
Mines of the High Desert .... Ronald Dean Miller
Mines of Death Valley .... L. Burr Belden
Mines of Eastern Sierra .... Mary DeDecker
Old Mines and Ghost Camps of California .... Ekman
Old California Mines .... Yale
Mines of the Mojave .... Ron and Peggy Miller
Ghost Towns of Arizona .... James & Barbara Sherman
Ghost Towns of New Mexico .... Jenkinson
Exploring the Ghost Town Desert .... Roberta Starry

LOST MINES, TREASURE HUNTING & METAL DETECTING

Treasure Hunter's Yearbook .... A.T. Evans
Getting Started in Treasure Hunting .... Alan Smith
The Treasure Hunter's Manual .... Karl Von Meuller
Map #1—*The Roadmap to California's Lost-Mines and Buried Treasures*
Map #2—*The Roadmap to California's Pioneer-Towns, Ghost-Towns, and Mining Camps*
The Modern Treasure Finder's Manual ...... George Sullivan
How to Use Detector Field Guide .... Roy Lagal
How to Test Detector Field Guide .... Roy Lagal
Electronic Detector Handbook .... E.S. LeGaye

Treasure Hunting with the Metal Detector .... Marvin & Helen Davis
Coinshooting—How and Where to Do It ...... H. Glenn Carson
Successful Coin Hunting .... Garrett
"X" Marks the Spot .... Goldberger
Thing: A Modern Search for Adventure ...... H. Glenn Carson
Backyard Treasure Hunting .... Lucie Lowery
An Eastern Thing .... Ralph Edmonds
Weekend Treasure Hunter .... Ryan
Weekend Desert Rat .... Ryan
Treasure Guide 1 .... Penfield
A Guide to Treasure in California .... Penfield
Famous Lost Mines of the Old West .... John Latham
Directory of Buried or Sunken Treasures and Lost Mines of the United States .... Penfield
A Guide to Treasure in Texas .... Thomas Penfield
A Guide to Treasure in Arkansas, Louisiana, Mississippi .... Thomas Penfield
A Guide To Treasure in Arizona .... Thomas Penfield
Lost Mines and Buried Treasures .... Mitchell
Lost Mines of the Great Southwest .... Mitchell
Canadian Treasure Trove .... Garnet Basque
Forgotten Mines and Treasures .... Wayne Winters
Campfires Along the Treasure Trail .... Wayne Winters
Guidebook to Lost Western Treasure ...... Robert Fergueson
Lost Desert Gold .... Ralph Caine
The Gold Hex .... Ken Marquiss
The Sterling Legend .... Estee Conaster
Pegleg—To Date and Beyond .. .. John Southworth

HISTORICAL

Here Rolled the Covered Wagons .. Albert & Jane Salisbury
Two Captains West .... Albert & Jane Salisbury
Old Forts of the Southwest .... Herbert M. Hart
Where the Washingtonians Lived .. .. Lucile McDonald
Early Oregon .... R.N. Preston
Forts of the West .... Frazer
Western Wagon Wheels .... Lambert Florin
Historic Western Churches .... Lambert Florin
Washington .... Bob & Ira Spring and Ron Fish
Historical Atlas of Oklahoma .... Morris
Historical Atlas of New Mexico .... Beck-Haase

Exploring Historic California ....Jack Adler
Odyssey of A Desert Prospector ....Albert
Western Mining ....Otic E. Young Jr.
The Mining Frontier ....Lewis
Mining Camp Days ....Emil W. Billeb
Comstock Mining and Miners ....Eliot Lord
An Empire of Silver ....Robert L. Brown
Gold Rushes and Mining Camps of the Early American West ......Vardis Fisher & Opal L. Holmes
The Nevada Desert ....Sessions S. Wheeler
Right-of-Way— A Guide to Abandoned Railroads in the United States ....Waldo Neilsen
Gold Cities ....Jim Morley and Doris Foley
Gold Mines of California ....Jack R. Wagner
Exploring California Folklore ....Russ Leadabrand
Travelers' Guide to the Comstock Lode ......Arthur Lassagne
Early California (Southern & Northern Editions) ..............R.N. Preston

TRAVEL GUIDE

Exploring California Byways, #2............Russ Leadabrand
Exploring California Byways, #3............Russ Leadabrand
Exploring California Byways, #4............Russ Leadabrand
Exploring California Byways, #5............Russ Leadabrand
Exploring California Byways, #6............Russ Leadabrand
Exploring California Byways, #7............Russ Leadabrand
Exploring Small Towns, No. 1: Southern California...................David Yeadon
Guidebook to Missions of California......Marjorie Camphouse
A Guidebook to the San Gabriel Mountains of California ....Russ Leadabrand
A Guidebook to the Southern Sierra Nevada..........................Russ Leadabrand
A Guidebook to the San Jacinto Mountains ..Russ Leadabrand
A Guide book to the Mountains of San Diego and Orange Counties.....................Russ Leadabrand
A Guidebook to the San Bernardino Mountains of California ....Russ Leadabrand
Where to Take Your Children In Southern California ....Davis Dutton & Tedi Pilgreen
A Guidebook to the Northern California Coast, Vol. I ....Mike Hayden

A Guidebook to the Northern California Coast, Vol II .... Mike Hayden
A Guidebook to the Mojave Desert of California .... Russ Leadabrand
Guidebook to the Feather River Country .... Jim Martin
Exploring Big Sur, Monterey and Carmel ...... Maxine Knox & Mary Rodriguez
Inside Death Valley .... Gebhardt
Exploring Death Valley .... Ruth Kirk
Ghosts of the Gold Rush .... George Koening
Exploring the Mother Lode Country .... Richard Dillon
Gold Rush Country ...... Editors of Sunset Books & Magazine
California State Parks .... John Robinson & Robert Calais
Southern California
Northern California
Back Roads of California .... Earl Thollander
Close Ups of High Sierras .... Norman Clyde
Death Valley Jeep Trails .... Roger Mitchell
Eastern Sierra Jeep Trails .... Roger Mitchell
Inyo-Mono Jeep Trains .... Roger Mitchell
Western Nevada Jeep Trails .... Roger Mitchell
Unbeaten Paths .... Ron Fish
Grand Canyon Treks .... Dr. Harvey Butchart
Alaska .... Bob & Ira Spring and Ron Fish
Byroads of Baja .... Walt Wheelock
Camping and Climbing in Baja .... John W. Roberson
Baja California .... Choral Pepper
Beaches of Sonora .... Walt Wheelock

CAMPING, OUTDOOR CRAFT

Nature and the Camper .... Mary V. and A. William Hood
Backpack Cookery .... Ruth Dyar Mendenhall
Backpack Techniques .... Ruth Dyar Mendenhall
Backpacking Made Easy .... Michael Abel
Basic Rockcraft .... Royal Robbins
A Camper's Desert .... Betty Tucker
The Golden Guide to Camping .... Robert E. Smallman
You Can Survive in the Outdoors .... Wayne Winters
To Hell On Wheels .... Alan H. Siebert
Unarmed Survival in Wilderness Alaska .. Michael A. Ganger
The Golden Guide to Scuba Diving .... Wheeler J. North

BOTTLES, RELICS & ARTIFACTS

Old Time Bottles .... Lynn Blumenstein
Redigging the West .... Lynn Blumenstein
Bottle Rush USA .... Lynn Blumenstein
Ghost Town Bottle Pricing Guide .... Wes & Ruby Bressie
Bottle Collectors's Handbook and Pricing Guide .... Yount
Old Bottles: How & Where to Find Them .... Abbott
Pocket Field Guide for the Bottle Digger .... Marvin & Helen Davis
Antique Bottle Identification Made Easy .... Robinson
Milk Bottle Manual .... Gordon A. Taylor
Ghost Town Relics .... Wes and Ruby Bressie
Treasure Hunter's Relic Identification .... Lynn Blumenstein
Truly America .... Lynn Blumenstein
What's It Worth and Where to Sell It .... Mack
The Golden Guide to American Antiques .... Ann Kilborn Cole
Treasury of Frontier Relics .... Les Beitz
Relics of the Redman .... Marvin & Helen Davis
Relics of the Whiteman .... Marvin D. Davis
Barbed Wire Handbook & Pricing Guide .... Turner
Barbs, Prongs, Points, Prickers & Stickers .... Clifton

COINS & CURRENCY

Guide Book of U.S. Coins .... R.S. Yeoman
Modern United States Currency .... Neil Shafer
Current Coins of the World .... R.S. Yeoman
Modern World of Coins .... Yeoman
Buying and Selling U.S. Coins .... Ken Bressett

Appendix C
An Advanced Project

For the more experienced treasure hunter-builder—one who is not timid about constructing a more complicated unit—the TAB editors are providing a bonus metal detector circuit. This super-sensitive unit will locate metal objects at greater depths than a metal locator of the beat-frequency type previously described. In this instrument, two separate boxes are required to house the circuits—one for a simple transmitter and one for a receiver—both of which use a loop coil in their tuned circuit. In the completed assembly, the coils are mechanically mounted 90 degrees from each other, i.e., the transmitter coil is in a vertical plane and the receiver coil is in a horizontal plane. Electrically, the radio frequency signal from the transmitter is at a null (minimum) as coupled to the receiver. Therefore, the receiver's meter will read zero. However, if the metal detector's coils are brought into proximitiy of a metal object, the transmitter-receiver signal path is changed slightly and will cause the meter to read upward. The larger or closer the metal object, the higher the meter reading. Frequency of operation is about 180 kHz.

CONSTRUCTION

Construction of the transmitter and receiver circuit boards is as for the simpler detector circuits described previously (on perfboards) so the same general techniques

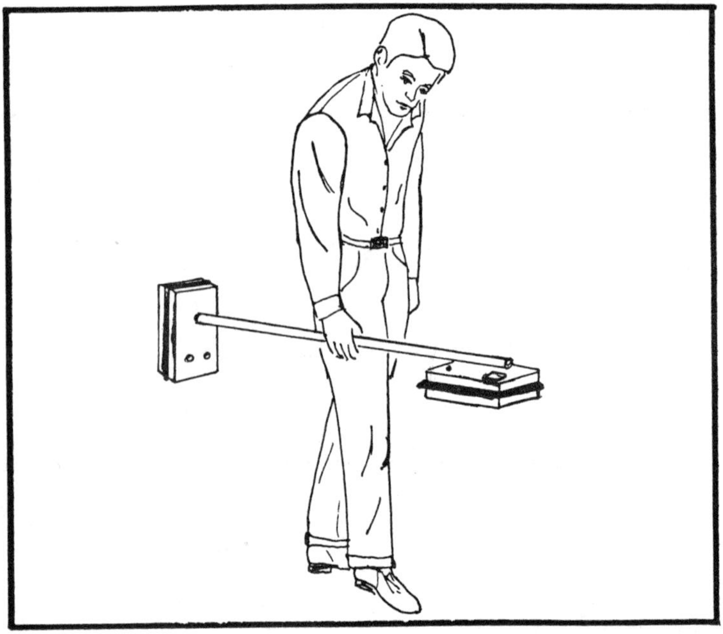

Fig. C-1. In this transmitter-receiver metal detector, the vertical loop behind the user is connected to a transmitter and the horizontal loop in front is connected to a receiver.

apply. When completed, the boards are each mounted into a plastic case. Now, one basic difference will be noted—the coils for each circuit are wound around the outside of each unit's plastic case. In other words, the case is utilized as the coil form as well as the housing for each circuit board. The transmitter and receiver are mounted at opposite ends of a 30-inch wooden pole or aluminum tube. Case is approximately 6 × 7 × 2 inches.

OPERATIONAL CHECK

Place the battery switch of the transmitter unit to the ON position. Tune a nearby AM broadcast radio until a steady buzzing sound is received. Turn the transmitter switch on and off to verify that the signal being received is from the metal detector circuit. Next, place the battery switch of the metal detector receiver circuit to the ON position. The receiver meter needle should move upward and a tone should be heard in the unit's headphones. Slowly adjust capacitor C1 on the transmitter for a maximum reading on the receiver meter.

PARTS LIST FOR TRANSMITTER

B1—9-V transistor battery
C1—365 pF variable capacitor
C2—3300 pF mica capacitor
C3—10,000 pF mica capacitor
C4, C5—0.1 μF, 200-V tubular capacitor
L1—34 turns of 22 enamel wire closewound outside of case
Q1—2N2924 transistor (Radio Shack 276-2016)
Q2—2N2646 unijunction transistor (Radio Shack 276-2029)
R1—150K, 1/2 watt carbon resistor
R2—39K, 1/2 watt carbon resistor
R3—1K, 1/2 watt carbon resistor
R4—180 ohm, 1/2 watt carbon resistor
R5—47K, 1/2 watt carbon resistor
S1—SPST miniature toggle switch
1—6 13/16 × 5 9/32 × 2 5/16-in. plastic case
1—6 1/2 × 5-in. cover
1—2 7/8 × 7/8-in. perfboard
1—48-in. aluminum conduit
1—Battery holder

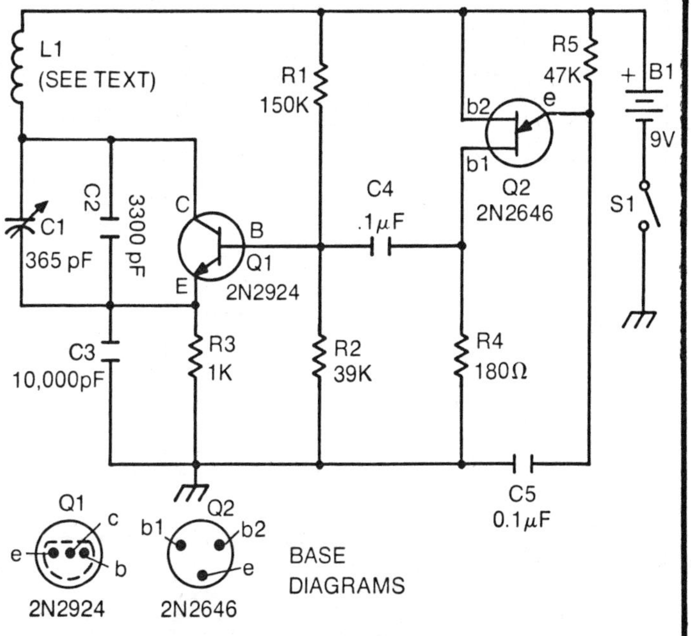

Fig. C-2. Transmitter and Parts List.

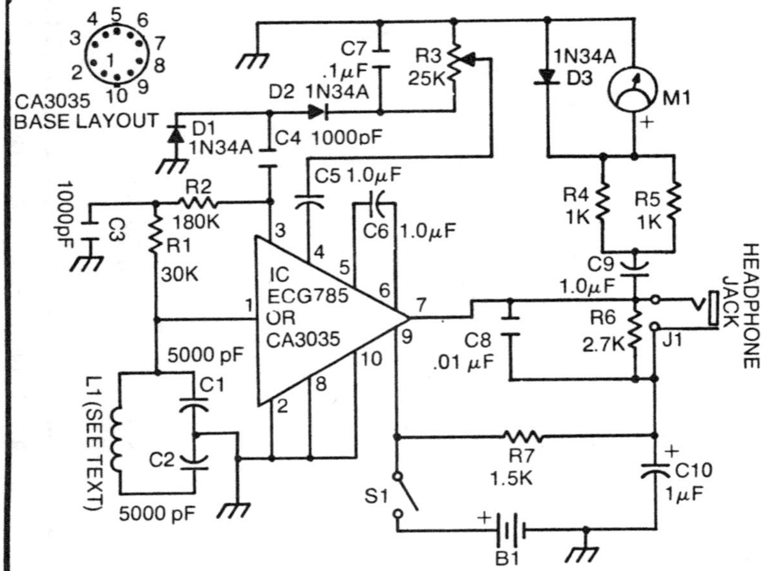

PARTS LIST FOR RECEIVER

B1—9 volt transistor battery
C1, C2—5000 pF mica capacitor
C3, C4—1000 pF disc ceramic capacitor
C5, C6, C9, C10—1.0 μF, 50-V electrolytic capacitor
C7—0.1 μF, 200V tubular capacitor
C8—0.01 μF disc ceramic capacitor
D1, D2, D3—1N34A diode (Radio Shack 276-1102)
IC—ECG785 integrated circuit (Sylvania) or CA3035 (RCA)
J1—Open circuit phone jack
L1—34 turns of #22 enameled wire closewound outside case
M1—500 μA DC meter
R1—30K, 1/2-watt carbon resistor
R2—180K, 1/2-watt carbon resistor
R3—25K, linear taper potentiometer
R4, R5—1K, 1/2-watt carbon resistor
R6—2.7K, 1/2-carbon resistor
R7—1.5K, 1/2-watt carbon resistor
S1—Spst miniature switch
1—6 13/16 × 5 9/32 × 2 5/16-in. plastic case
1—6 1/2 × 5-in. cover
1—2 7/8 × 2 7/8-in. perfboard
1—Battery holder

Fig. C-3. Receiver and Parts List.

This tunes the transmitter to the receiver frequency. Mechanically adjust the relative position of the receiver to the transmitter on the supporting pole or tube to obtain a minimum meter reading. This must be done while as far away from any metal objects as possible. The unit should now be peaked for operation.

Index

Index

A

Adjustment, control 85
Air sensitivity test 97, 99
Artifacts, Civil War 56

B

Basing diagram, transistor 59
Batteries as cause of malfunctions 89, 92
Beat frequency 27
    oscillator 78, 79
    oscillator detector, (BFO), use of 88
Bending transistor leads 64

C

Capacitance
    parallel 68, 70
    series 68, 70
Capacitors, types 58
Center-tapped search coil 75, 77
Chart, troubleshooting 95
Circuits
    detector 67
    meter 71, 73
Civil War artifacts 56
Coding, resistor 24, 25
Coil
    search, center-tapped 75, 77
    search, scramble-wound 31, 33, 46, 47
    search, winding 46, 47
Component
    layout 61
    substitute 67, 70
Connections, solder, faulty 93
Control
    adjustment 85
    shaft 38, 39

D

Detector
    beat frequency oscillator (BFO) 78, 79
    beat frequency oscillator (BFO), uses of 88
    circuits 67
    circuit schematic 46, 72, 73, 74, 76, 78
    maintenance 89, 96
    operation 41, 55, 56, 83, 86
    repairs 89, 96
    schematic diagram 62, 63
    sensitive 75
    testing 40, 54, 55, 64, 65, 82, 83
    transmitter-receiver (TR) 78, 79
    transmitter-receiver (TR), uses of 88

Diagram, schematic, detector 62, 63
Diodes 58
Diode polarity 60
Digging your find 100

E

Electronic symbols 18, 19
Erratic operation 91, 94

F

False signal caused by grass 55
Faulty solder connections 93
Field operation 99, 100
Find, digging your 100
Fisher, Gerhard 11
Frequency
  beat 27
  instability 90, 91
Frisbee as search coil housing 28, 37

G

Grass, cause of false signal 55

H

Heathkit Model GD-348 81, 88
  Deluxe Metal Locator

I

Indicating meter 71, 73, 77
  78, 84
Instability, frequency 90, 91

K

Kits, detector 81, 88

L

Layout, component 61
Location, search head 84

M

Maintenance, detector 89, 96
Malfunctions, 89, 92
  batteries as cause of
Meter
  circuit 71, 73
  circuit schematic 78
  indicating 71, 73, 77
  78, 84
Monitor, portable radio as 43, 71

N

NPN transistor 59

O

Operation
  detector 41, 55, 56
  field 99, 100
  erratic 91, 94
  detector 83, 86
Oscillator, beat frequency 78, 79

P

Paper plates as housing 47, 48, 49
Parallel
  capacitance 68, 70
  resistance 68, 70
Plates, paper, as housing 47, 48, 49
PNP transistor 59
Poor sensitivity 90
Polarity, diode 60
Portable radio as monitor 43, 71
Private property,
  searching on 101, 102
Procedures, tuning 97, 99
Property
  private, searching on 101, 102
  public, searching on 101, 102
Public property, searching on 101, 102

R

Radio, portable, as monitor 43, 71
Reading schematics 18
Repairs, detector 89, 96
Resistance
  parallel 68, 70
  series 68, 70
Resistor
  coding 24, 25
  values 24, 25

S

Search
  coil, center-tapped 75, 77
  coil, Frisbee as housing 28, 37
  coil, scramble-wound 31, 33, 46, 47
  coil winding 46, 47

head location 84
head submersible 81, 82
Searching
on public property 101, 102
on private property 101, 102
Sensitive detector 75
Sensitivity
poor 90
test, air 97, 99
Series
capacitance 68, 70
resistance 68, 70
Scanning speed 99
Schematic
detector circuit 46, 72, 73 74, 78, 76
diagram, detector 62, 63
meter circuit 78
reading 18
Scramble-wound search coil 31, 33, 46, 47
Shaft, control 38, 39
Signal, false, caused by grass 55
Solder connections, faulty 93
Soldering 15, 16
Speaker volume 89, 90
Speed, scanning 99
Submersible search head 81, 82
Subsitute components 67, 70
Symbols, electronic 18, 19

T

Techniques, soldering 16, 23, 24, 34
Test
equipment for troubleshooting 94, 96
sensitivity, air 97, 99
Testing
detector 40, 54, 55, 64 65, 82, 83
transistors 94, 96
Tools 12
Transistors 58
basing diagram 59
leads, bending 64
NPN 59
PNP 59
testing 94, 96
types 59
Transmitter
receiver detector 78, 79
receiver (TR) detector, uses of 88
Trimmer, mounting 35, 36
Troubleshooting 94, 96
chart 95
test equipment for 94, 96
Tuning procedures 97, 99

U

Uses
of beat frequency oscillator detector 88
of transmitter-receiver (TR) detector 88

V

Volume, speaker 89, 90

W

Workbench 12